# MATTERPHORICS

# MATTERPHORICS

*On the Laws of Theory*   DANIELA GANDORFER

Duke University Press   *Durham and London*   2026

© 2026 DUKE UNIVERSITY PRESS. All rights reserved
Cover design by Dave Rainey
Typeset in Garamond Premier Pro by Westchester Publishing Services

Library of Congress Cataloging-in-Publication Data
Names: Gandorfer, Daniela author
Title: Matterphorics : on the laws of theory / Daniela Gandorfer.
Description: Durham : Duke University Press, 2026. | Includes
    bibliographical references and index.
Identifiers: LCCN 2025034051 (print)
LCCN 2025034052 (ebook)
ISBN 9781478032953 paperback
ISBN 9781478029472 hardcover
ISBN 9781478061687 ebook
Subjects: LCSH: Law—Philosophy | Technology and law | Process
    philosophy | Matter—Philosophy
Classification: LCC K230.G3625 A28 2026 (print) |
    LCC K230.G3625 (ebook) | DDC 340/.1—dc23/eng/20251119
LC record available at https://lccn.loc.gov/2025034051
LC ebook record available at https://lccn.loc.gov/2025034052

Cover art: Visualization of the author's breath, measured with a
customized sensing device and minted as a non-fungible token.
Part of A Thousand Breaths, a global collection developed by LoPh+
within the *Decentralized Right to Breathe* initiative, materializing
breath as participatory, distributed, and nonproprietary normativity.

# Contents

# Acknowledgments

Everything matters, although differently, which is also to say that we are accountable for more than the words we put onto paper, and we owe our thoughts to more than the names we remember. Acknowledgments—expressed in words, written and placed at the beginning or the end and yet coming from the middle—cannot do justice to what it means to write and think collaboratively: the impact it has, and will have, on us, on others, and on books printed as well as books imagined but, in any case, existent. And yet I wish to thank those most present and influential for this book project. For it is not in the least a figure of speech when I say that this book would not have been possible without those who generously thought with me, those with whom I was able to think, and those not willing to think along the same lines yet accepting, often even encouraging, of disagreement and, most importantly, difference.

I am deeply grateful to Jonathan C. Aguirre, Karen Barad, Jane Bennett, Rosi Braidotti, Judith Butler, Eduardo Cadava, Emanuele Coccia, Marianne Constable, Maria Drakopoulou, Raviv Ganchrow, David Gissen, Deniz Göktürk, Suzanne Guerlac, Allen Feldman, Niklaus Largier, Andrea Leiter, Achille Mbembe, Jack Halberstam, Bernard Harcourt, Stefan Helmreich, Hyo Yoon Kang, Sara Kendall, Sanford Kwinter, Debarati Sanyal, Marie Petersmann, Luke Mason, Andreas Philippopoulos-Mihalopoulos, Dimitri Van Den Meerssche, Elizabeth Povinelli, Brendan Rogers, Nofar Sheffi, Massimiliano Tomba, Tanja Traxler, Jesús Velasco, Eyal Weizman, Patricia J. Williams, Carlos Yanes, and Carey Young for our conversations and for their comments, critique, laws, waves, time, sounds, letters, drawings, images, maps, NFTs, imaginations, invitations, books and manifestos, ETHS, concepts, meals, tunes, signatures and letters, and encouragements.

My heartfelt thanks go out to all the extraordinary LoPh[+] researchers, Web3 coders, designers, engineers, activists, lawyers, scientists, communities, and avatars who have guided me in thinking-with and breathing-with matter(s) on the ground.

The bittersweet madness of writing and publishing a book was possible—peculiarly pleasurable, accompanied by tearful laughter and scarf-ed pressure, moves and stillstands, and synesthetic notes, which, so I hope, eventually will find themselves well—because of the relentless support of Peter Goodrich and Zulaikha Ayub, the unwavering trust of Dejan Ivković, the touching care of friends and more-than-friends, the support of my family, and the presence of little rays of sunshine, matterphorically, both those already on this planet and those yet to come (or endlessly imagined): Maja J., Klara C., Mia I., Julius T., Adrian, and Lilith.

# Introduction: How to Begin?

A GUIDE

The library contains not books
but glaciers.
The glaciers are upright.

Silent.
As perfectly ordered books would be.
But they are melted.

What would it be like
to live in a library
of melted books?

With sentences streaming all over the floor
and all the punctuation
settled to the bottom as a residue.

It would be confusing.
Unforgivable.
An adventure.
—ANNE CARSON, "Wildly Constant"

What men are poets who can speak of Jupiter if he were like a man, but if he is an immense spinning sphere of methane and ammonia must be silent? —RICHARD FEYNMAN, *The Feynman Lectures on Physics*

And I, I am feeling a little peculiar.
—4 NON BLONDES, "What's Up"

In the beginning of *Cannibal Metaphysics*, Eduardo Viveiros de Castro reveals that he had originally planned to write a book as an homage to Gilles Deleuze and Félix Guattari. He would have written it from the point of view of his discipline, anthropology, and he would have called it *Anti-Narcissus: Anthropology as Minor Science*. Somehow, however, he sensed that this endeavor might be too unsettling, too contradictory, perhaps even too daring to be undertaken by him, especially alone. Instead of writing the book himself, he decided to write about it "as if others had written it." He decided to write a beginner's guide to the "endlessly imagined" book, which "ended up not existing—unless in the pages to follow."[1] I found myself intrigued by this "invisible" book: written by not only one but more than one, an homage to Deleuze and Guattari's work and existing only *in* and *as* relations. Viveiros de Castro's beginning, his introduction, continues with an outlining of what is at stake—not yet with *Cannibal Metaphysics*, but with *Anti-Narcissus*, of which a draft, despite being fictional and invisible, "has begun to be completed by certain anthropologists who are responsible for a profound renewal of the discipline."[2] The coming into existence of that book, whatever its form(s), is not only a collaborative endeavor addressing the stakes of a discipline; like the beginner's guide, it is also a political project that aims to make anthropology the theory and practice of the "permanent decolonization of thought."[3] Viveiros de Castro's beginning comes from the middle, an active political project within and beyond institutional frameworks, carrying an ethics of relationality and collaborative thought that harkens back not only to Deleuze and Guattari's work but also to the thoughts that made their thoughts possible and the works with which their works were and will be written.

Despite *Cannibal Metaphysics* bearing Viveiros de Castro's name, declaring him the author is, to adopt the phrase Deleuze and Guattari use to describe their cowriting of *Anti-Oedipus*, "already quite a crowd."[4] Deleuze, both present and absent in the three parts to come, insists that there is a major difference between an author and a writer. The writer does not assume a world that awaits us to be created but rather "invents assemblages starting from assemblages which have invented him."[5] One must, he continues, write *with* the world, *with* a part of the world, *with* people, not in place *of* but *with*.[6] Put differently, writing is not the skillful practice of an isolated genius expressing truth in language but rather an encounter that unfolds as a becoming. In writing, "one always gives writing to those who do not have it" (without writing *for* them or in their name), while those who don't have it "give writing a becoming without which it would not exist" or would simply "be pure redundancy in the service of the powers that be."[7] Because writing, for Deleuze, cannot be in the service of power, there is also no other reason for writing than "to be a traitor to one's own reign," to one's majority, and to writing itself.[8] It is not easy to be, let alone to write as, a traitor—a fact that some writers know better than others. When Baruch Spinoza, the philosopher whose mode of thought and ethics Deleuze held in the highest esteem, began to write the *Ethics* in the seventeenth century, he already thought from a precarious place: "while it sometimes happens that a philosopher ends up on trial," Deleuze writes, "rarely does a philosopher *begin* with an excommunication and an attempt on his life."[9] And when the Italian philosopher Antonio Negri, more than three hundred years after the publication of the *Ethics*, takes on Spinoza's work in his *Savage Anomaly: The Power of Spinoza's Metaphysics and Politics*, he does so while coming to terms with the fact that he too must begin writing from an arresting place. In his introduction, Negri discloses that "this work was written in prison" and that, once in prison, he "started from the beginning: reading and making notes" and asking his colleagues to send him books.[10] Neither Deleuze nor Viveiros de Castro have written from prison, and, to my knowledge, neither's life has been in immediate danger as a result of having written an academic and philosophical book. Yet both have written with a particular mode of thought that, as I will show, challenges dominant political and onto-epistemological assumptions,[11] making it inevitable to think closely about what it means not only to write but also to cowrite and write-*with*. This is as much a question of ethics as it is of politics. In thinking back on cowriting with Guattari, Deleuze states that what was important was less working together than "this strange fact of working between the two of us," creating a multiplicity; not only does one cease to be an author but "there is politics, micro-politics."[12] Assuming that this is true not only for *Anti-Oedipus*, *A Thousand Plateaus*, and *Anti-Narcissus* and its beginner's guide *Cannibal Metaphysics* but also for a mode of writing, thinking, and doing theory that is impersonal yet accountable, inherently collaborative yet singular, I have felt compelled to attend to the crowd I am and to the imagined book to which I want to contribute.

If I had written a book, it would have been one on law and its states of matter, arguing that law—every single law, in fact—is an expression or *configuration* of entangled physical, material, biological, chemical, political, cultural, and representational matters and forces. It would have drawn from, among other fields, legal theory, literature and media studies, and the sciences, and its title would have referenced *A Thousand Plateaus* rather than *Anti-Oedipus*. The political project carrying this imagined book would have been to facilitate a literacy and attentiveness for perceiving, sensing, and understanding the existence of law—not simply its effects and consequences but its entanglement with each and every particle and in each and every moment, especially in those of precarity and where questions of life and death, or death and non-life, are at stake. It would, for example, have exposed in a convincing manner how the drowning of an asylum seeker in the Mediterranean Sea is as much an outcome of political and ethical negligence, in large part deliberately orchestrated, as it is the result of the very constitution of particular laws, which cannot be disentangled from the materiality of the ocean, from Boyle's law, from molecular and atomic interactions, from forces of all kinds, or even from the physical limits to what a human body can do: how far it can swim, what temperatures it can endure and for how long, how lungs fill with water, and how breathing becomes impossible. Attending to incidents of injustice and to how both the matter and materiality of law are rendered nonexistent precisely because they allow for violence and injustice to continue taking (a) place would have, over the course of this book, brought about a mode of sensing law that requires a different ethics and a different sense of response-ability—a *synaesethics*.[13] This, of course, would have been too extensive and too delicate of a project for a single academic text to take on. It is a political project of this particular timespace, evolving amidst the rise of various forms of fascism and authoritarian regimes; ecological and environmental shifts resulting from global warming; global pandemics; the ensuing proliferation of military and social conflicts across the globe; neocolonial undertakings on earth and in outer space; scientific developments in, among other fields, quantum physics, synthetic biology, and nanotechnology; and new digital technologies, including blockchain, generative AI, virtual and mixed reality, and the Internet of Things (IoT). At this time—and in these times—the book on law and its states of matter is, I would argue, overdue. Fortunately, like *Anti-Narcissus*, this book has already begun to be written by those aiming to decolonize and materialize law and legal theory. As such, it aims to unthink modernist legal thought: its mind and matter dualism, its ubiquitous representationalism, its bounded idea of the subject, its predilection for transcendence, and the modes of power and governance it has created and rigorously defends. Many of the thinkers cited and referred to in this book have already been traitors to their disciplines, to their institutional frameworks, and to dominant modes of thought. They have thereby paved the way for those willing to think along with them to articulate arguments and theories that contribute to precisely this project. Despite knowing that this book will be, if successful, a beginner's guide to the project, making a case for its importance and timeliness, the question of *how* and *where* to begin remains to be addressed.

## Beginnings: Synaesethics of Thought

The complexity of this question is what attracted me to Viveiros de Castro's beginning, to the concept of a beginner's guide, and to the imagined *Anti-Narcissus*, which exists only by means of relations and in the pages to follow. While my research, teaching, and social entrepreneurship focus on emerging, tech-informed normativities in frontier spaces, I am also trained as a scholar in literary and media theory. My own curiosity about the power of Viveiros de Castro's gesture and its potential to unsettle the disciplines close to me leads me to start with the question of the medium. Indeed, I would argue that the beginner's guide and *Anti-Narcissus*—a virtual yet real book aiming to decolonize thought without taking the form of a book itself—urge us to question what a book in general, and an academic text in particular, *is* and can be: the kind of relations it consists of and that make it possible. What's more, they prompt us to rethink the practices of academic writing, thinking, and reading and the common notions we attribute to them. Thinking about *Anti-Narcissus*, how are we to understand the multiplicity of forms, relations, and entanglements that make up what we consider a book and yet that exceed it by far—especially if what is expressed by a book (or text) is not simply content, ready to be extracted, critiqued, affirmed, negated, cited, and redacted, and cannot be contained within a fixed form (that of a letter, word, sentence, page, or book)? And, regarding the beginner's guide, how are we supposed to read an academic text that is based on, refers to, quotes, and invokes a book that does not exist in any recognizable form but nonetheless "radicalize[s] the reconstitution of the discipline?"[14] Asking where and when something becomes a *book*, and when it ceases to be one, means confronting and navigating different forms of power: the power to determine beginnings, dominate, colonize, rule, judge, and speak truth on the one hand; the power to resist and destroy oppressive modes of thought, think-with more-than-friends, and withdraw from rather than take (a) place on the other. Given the long history of the book and its assumed relation to both (written) truth and law, this is not an idle question. It touches on the delicate matters of authority, representation, authenticity, originality, foundation, and law, all of which are crucial pillars of a dominant (and dominating) mode of thought. Consider, for example, that the academic discipline of German language and literature studies, fundamentally rooted in philology—the study of the history of language and its sources—emerged from legal science at the beginning of the nineteenth century. Indeed, the establishment of said academic field is attributed to Jacob Ludwig Karl Grimm, who, although commonly known as having co-edited *Grimms' Fairy Tales* with his brother Wilhelm Carl Grimm, was actually a legal historian, philologist, and student of the influential German legal scholar Friedrich Carl von Savigny. (The latter will reappear, for a fleeting moment, in the first part.) Besides literary and legal works, Jacob Grimm wrote and co-edited *foundational* books, such as the *Deutsche Grammatik* (German Grammar), the *Geschichte der Deutschen Sprache* (History of the German language), and the *Deutsche Wörterbuch* (German Dictionary), deliberately seeking in philology a beginning and origin narrative, not only of German studies but of a cultural and national German identity in formation. Needless to say, this too was a vast political and legal project. It is imperative to acknowledge that the power of linguistic formation and the existential role language plays in determining not only identity but existence have been pivotal tools of colonialism, a notion famously articulated by postcolonial thinker Frantz Fanon: "to speak is to exist absolutely for the other."[15] To speak, to use the language of power, and to obey its law is to begin to exist, to think, and to be thinkable.

Viveiros de Castro's *Cannibal Metaphysics* exposes, although differently, the predicament—but also the potential—of beginnings. By announcing to have written a "beginner's guide" to a book that exists in/as relation(s), Viveiros de Castro demonstrates that each beginning—of writing, of thinking, of a text and a book—unfolds from the middle, from a vast field of actual and virtual (im)possibilities and relations across space-time. *Anti-Narcissus* exists, and yet it might not have *a* beginning. It is powerful because it multiplies possibilities of what a book can be and do, yet it is not foundational. For Deleuze and Guattari, the concept of the middle is a crucial concept to think-with and think-from. It denotes neither an average nor a line that connects two points but rather a milieu, a space thick with relations "where things pick up speed" and becomings evolve.[16] As such, the middle is a *between* that does not denote a localizable relation but rather "a transversal movement that sweeps one and the other away, a stream without beginning or end that undermines its banks," that does away with foundations and nullifies endings.[17] Although this book is not directly concerned with questioning, addressing, or deconstructing beginnings, foundational moments, and spaces, it remains critical of any mode of thought that derives its force and power, and often violence, from a claimed origin mostly understood as immaterial and representational. As such, I am thinking-with Deleuze—in particular, with the third chapter of *Difference and Repetition*, which opens precisely with the question of beginnings, stating that there is, in fact, no "true beginning" of philosophy, for "beginning means eliminating all presuppositions": a noble aspiration, yet one that has not been achieved by Western philosophy, despite its constant reflections on that "very delicate problem."[18] Even Descartes's "I think, therefore I am," Deleuze points out, does not denote a true beginning, as it already assumes everything that there is to assume: that "everyone knows, independently of concepts, what is meant by self, thinking, and being."[19]

Part I takes its cue from this delicate problem and argues for a mode of theory that starts from the middle and refuses to claim territory or even (a) place: *matterphorics*. Highlighting both the pitfalls of representational modes of thinking and the ethical imperative to redefine what it means to think, the section challenges established notions of knowledge production and institutional power. Using the neologism *synaesethics*, it advocates for a mode of thought guided by an ethics of collaborative sensing and sensemaking, attentive to the inseparability of mind and matter and the response-ability that there arises. Indicating the collaborative nature of thinking (*syn-*) and the nonrepresentational practices of sensemaking, the term also refers to *synesthesia*—a phenomenon that challenges traditional neurology by highlighting the complexity, interconnectedness, and plasticity of sensory processing in the brain while also questioning humanist ideas of the separation between mind and matter, consciousness, and representationalism. Synaesethics, committed to the matter(s) of thought and a mode of theory that matters (matterphorics), bears the potential for different modes of existence to become thinkable and livable. The section demonstrates that both thought and theory must lose their minds, become unreasonable, commit the capital crime of treason, break the image of thought, question metaphor, and engender thinking in thought in order to become response-able and sensitive to what is not yet thinkable, to what at first might seem unthinkable and nonsensical: an electron crashing into language, a Red Bull logo breaking through the sonic barrier, oxygen molecules pushing legal theory, life and law becoming cutting edge.

## From the Middle: Matterphorics

Precisely because this book questions assumptions about beginnings, foundations, and origins—to test what theory can do if it rejects the Cartesian dualism of matter and mind, the inherently hierarchical separability of being and thinking—it tends, as a matter of method and resistance, to unfold from the middle: Donna Haraway calling out Bruno Latour for thinking-with the Nazi legal scholar Carl Schmitt, who embraced colonial land appropriation as paradigmatic beginnings; the Austrian parachutist Felix Baumgartner, standing at the edge of a balloon gondola about to free fall from the stratosphere while openly calling for Europe to close its borders to refugees and asylum seekers; a scalpel cutting through the body of conjoined twins, attributing a right to (a) life to one of the resulting bodies while cutting the other out of life; strong nuclear force overpowering electromagnetic force, leading not only to the fission of an atomic nucleus but to a chain reaction that bears the potential to erase and create space-time; a decentralized right to breathe emerging from a thousand plateaus calling for "another justice, another movement, another space-time," a tune yet to be picked up by a mode of theory that can attend to what comes from the middle.[20] All these cases—matterphorical case studies—traverse disciplinary boundaries, expose complicities, and trace entanglements, revealing that what underlies concepts, including legal ones, is not (epistemological) uncertainty—conveniently resolved by the rational subject and detached from the world's matter(s)—but an indeterminacy that disregards the separability of mind and matter. This shifts the issue from one of knowability to one of "what can be said to simultaneously exist" and in turn necessitates a different mode of doing theory and creating concepts: *matterphorics*.[21]

Part II reveals the import of a matterphorics of law by presenting a matterphorical case study of a man falling from the sky. Here, we uncover the inherent indeterminacy within legal concepts and underscore the ways they matterforth. In doing so, the section examines the entanglement of law with matter and force(s), challenging the hegemonic understanding of law—and legal concepts—as ideational, representational, and detached from the physical, embodied world. The matterphorical case study exposes the scope of concepts, taking the reader not only on a journey but also for a fall, ride, and dive—from space to the deepest point in earth's oceans, from energy drinks to drownings, from sovereign to meme dog, from liquid oxygen to held breaths, from canons to cannons, from ballads to ballistics, from literature to law (and vice versa), from vertigos to vortices, from kings and territories to waves and forcefields, from naked ideas in a tub to Felix Baumgartner's pressure helmet. Each movement eventually reveals the indeterminacy of even the most powerful concepts: border, legal regime, territory, legal subject, rights. It is no coincidence that the space that reveals this indeterminacy most unapologetically is that of the frontier. Indeed, as the section shows, the frontier—a "shifting terrain between legality and illegality . . . violence and law, restoration and extermination"[22]—is where both indeterminacy and the desire to determine, regulate, appropriate, and territorialize are most pronounced and palpable. What's more, frontiers—be they outer space, the Arctic, the Antarctic, the deep sea, metaverses, or the crypto-space—highlight the fluidity and malleability of legal concepts, which are subject to forces of all kinds. *Matterphorics* emphasizes the importance of acknowledging the materiality and complexity of these spaces and advocates for an ethical and participatory approach to legal sensemaking that is response-able to the shifting realities of frontier spaces. This also reveals a crucial underlying claim of the book—namely, that doing theory matterphorically, and engaging in an ethics of sensing and sensemaking (synaesethics), does not imply avoiding power. Rather, as Brian Massumi states, it demands creatively getting "down and dirty in the field of play," mobilizing complicity toward new kinds of emergences, staying with the trouble, and adamantly refusing to believe that "the game is over" or that there is "no sense having any trust in each other in working and playing for a resurgent world."[23]

Part III returns to the beginning, Haraway's critique of Latour and the Anthropocene as earth's life story, to take a closer look both at the construction of the concept of *life*, an inherently ambiguous and indeterminate concept, across disciplines and what it means to have, take, and narrate (a) life. Given the progression of global warming and its differentially palpable effects, whose scale and complexity exceed borders, cultures, and modes of being, the question of who or what has the privilege to form a life story ought to be raised: What is the role of law in creating and sustaining this story, what does it have to do with genre, what concept of life underlies it, why does it matter (and for whom or what), and what can theory do? The section approaches these questions by tracing the complicity between not only life and legal theory—looking closer at what it means to have a right to (a) life—but also life and literary theory. Showing what it means for legal theory to become cutting edge—response-able to the cuts it performs and for questions of mattering—the section offers a matterphorical case study of the legal subject/person as bearer of rights, cautioning that, despite its importance, the concept operates through representational thought and the performance of onto-epistemological cuts. By demonstrating how the concept of autobiography, as the writing of the self's binary *life story*, finds its expression in the narration of earth stories, this section advocates for a more-than-real approach to literary and legal theory in alignment with a *jurisliterary* perspective on law, literature, and life.[24]

## The End? A Beginner's Guide to Emerging Normativities

I started writing this book many years ago as a beginner's guide to the book on law and states of matter: an endlessly imagined book, written by more than one, existing only in the pages that follow—only *in* and *as* relations. It became evident rather soon that this book not only subverts the authoritative power of beginnings but, to stay committed to a synaesethics of thought, must also resist conclusions, determinations, and consensus. It is not only the power of the first word but also the finality of the last—whether conclusion or verdict—that reiterates the illusion of thought reaching a self-satisfied consensus. A book (or guide) on how to do theory matterphorically, how to create (legal) concepts that matter-*forth* differently, a book that argues for going beyond the limits of critique and experimenting creatively with complicity to make different modes of existence possible, cannot speak a final word, whether confidently or by following academic convention. If trained as an academic, how does one do theory from the middle, on the ground? How does one, as Massumi calls for, mobilize complicity toward new kinds of emergence?

Engaging with these inquiries prompts us to consider what it means to start doing theory where language and representational systems more broadly reach their limits. What does it take (quite literally) to create concepts that can withstand the pressure of new digital technologies, resist colonial and proprietary desires, and adapt to the shifting material conditions on earth? These concepts must emerge not as universal forms but as matterphorical expressions of a particular space-time: situated, understood in their specificity and historicity, and with awareness of the power they carry to matter-forth. What is the role of the critical scholar and the university? And importantly, what if concepts were to be created collaboratively, where the thinker thinks but does not hold knowledge exclusively, contributes but does not author, engenders but does not direct, establishes relations but does not determine the relata? What is the cost of democratized, participatory concepts; where do the funds come from; how do we navigate complicity with states and markets; and where do these concepts draw their power from if not centralized institutions, international contracts, supply-and-demand dynamics, or private wealth?

Grateful for the education I have received and the spaces the university provides for me yet acutely aware of its exclusivity and power—as well as its affinity for modes of thought that protect both—I resolved to seek alternative vehicles for thought. Recognizing that theory is always already in the world and witnessing the profound impact it can have on questions of law and justice through academia-borne entities such as Forensic Architecture, I, together with colleagues Zulaikha Ayub and Jonathan C. Aguirre, cofounded LoPh⁺ with seed money from Princeton University. Now a multi-award-winning nonprofit organization registered in Vienna, Austria, LoPh+ works on decentralized justice and tech-enhanced, community-based governance models, seeking different kinds of legal concepts for a more inclusive and diverse future to emerge. The lab's most prominent initiative—De.RtB (Decentralized Right to Breathe)—is an experiment in creating a radically different, participatory, and decentralized concept of rights. The final section of the book thus indicates what *Matterphorics* functions *with*—not only within academic confines but also beyond its walls, structures, logic, and immediate reach—including a continuously materializing decentralized right to breathe: a matterphorical legal concept in the making, created by a crowd of more-than-friends, decentralized in its power, distributive in its nature, participatory in its constitution, and affirmative in its operation. The commitment to infuse multiplicity into emerging technological frontier spaces, engaging with and diversifying these domains, is rooted in my extensive experience in these fields. Years of working and researching within these spaces illuminated the necessity for alternative, nonrepresentational, and radically relational approaches. In frontier spaces in particular, including the new governance frontier where emerging normativities are negotiated, claimed, and created, these approaches must be conceived in concert with new digital technologies. This includes distributed ledger technologies, like blockchain, decentralized AI, and the IoT, in order to ensure that they can endure shifting power dynamics and mounting pressures. Concepts, too, need to be powered, and what a technology can *do* depends on who has access to its creation and use. We need crowds, not geniuses. Besides, crowds are smarter anyway. Let us start, again yet not all over, together-apart by matterphorically going all the way down, becoming impersonal, creatively gaming complicity, acting response-ably, betraying matterphorically, desiring unreasonably, falling synaesethically, and failing successfully.

# I

# Synaesethics of Thought

*Doing Theory Matterphorically*

I

# Thinking-With Matter(s)

UNLOVING THOUGHT,

UNFRIENDING THEORY

But, of course, these theorists do not come and engage my friends—nor do they seem to think they must—given that philosophical thought defines itself as a kind of thinking that can generate thought for all beings without engaging most; and all truths remain the same no matter where you perceive them. Does their disinterest matter? —ELIZABETH POVINELLI, *Geontologies*

It matters what thoughts think thoughts. It matters what knowledges know knowledges. It matters what relations relate relations. —DONNA HARAWAY, *Staying with the Trouble*

Who speaks and who acts? It's always a multiplicity, even in the person that speaks or acts. We are all groupuscles. There is no more representation. There is only action, the action of theory, the action of praxis, in the relations of relays and networks. —GILLES DELEUZE, "Intellectuals and Power"

In *Staying with the Trouble*, Donna Haraway criticizes Bruno Latour, one of her dearest academic friends, by arguing that the thought he chooses to think-with is inhospitable to the earth and its many possible futures. Latour draws on Carl Schmitt's concept of the enemy, "with all its tones of host, hostage, guest, and worthy enemy," and, although Latour makes clear he does not want to draw upon Schmitt's story, allows a narrative of "trials of strength" and "mortal combat" within which "the knowledge of how to murder each other remains well entrenched."[1] In the shadows of the possibility of this atrocious violence lies not only the enemy but the complimentary concept of the *friend*: the brother in arms to whom we owe our lives and to whom we owe it to take lives deemed hostile. Asked about this passage in *Staying with the Trouble*, Haraway states that "Bruno is someone, as a friend and thinker, I love," and she explains that her engagement with his thought is not "in the mood of destructive critique, but something else."[2] Political theorists, Haraway continues, are always writing about speculative fabulations, with only some, including Schmitt, becoming canonized as such. However, Haraway turns to feminist speculative fiction writers "from many communities and of several genders"—among others, N.K. Jemisin, Rebecca Roanhorse, Nancy Jane Moore, and Neon Yang (formerly JY Yang)—stating that "they do political theory in a way Carl Schmitt was never capable of, and I want my friends, like Bruno, to do it that way."[3] I wish to begin this chapter on what it means to think and think-with precisely with the predicament of critique and friendship because it emphasizes that the question of whom to think-with is not simply one of personal preference; it is a question of ethics, demanding thought to push beyond its image, beyond the constraints and comfort of what registers as known and familiar. At the same time, this predicament draws attention to matters of complicity and loyalty, serving as a reminder that a departure from the familiar and established, even if viewed as a form of critique, remains situated within fields of power. In his *Postcapitalist Manifesto*, Brian Massumi importantly cautions that complicity, understood in relation to capitalist processes, "is an ontological condition under neoliberalism."[4] Recognizing the fact that there is no being outside, not least because there is "no way of surviving without being complicit with it," it is necessary not only to acknowledge that critique is not enough but also to "experiment with modulating complicity by learning how to inflect it toward other kinds of emergences."[5] To engage in "creative duplicity," as Massumi suggests, means gaming complicity with what remains unacknowledged, unrelated, denied, and even, as in Massumi's argument, declared the enemy: neoliberal capitalism.

Related to the predicament of academic friendship—this intricate knot of personal relationships, academic hierarchies often tied to letters of recommendation, and the bittersweet illusion of having to carry on the torch—and the limits of critique is the question of loyalty. Derived from the Latin *lex*, meaning law, loyalty carries within its negation treason: the capital crime, the killing of the sovereign, the overthrow of the government. Loyalty is a rather serious matter. Deleuze already highlights the stakes of the complicity of thought by distinguishing the *traitor*, who turns against "the world of dominant signification," established order, state thought, and signifier, from the *trickster*, who "claims to take possession of fixed properties, or to conquer a territory, or even to introduce a new order."[6] Being a traitor means to create and experiment *in* the world, yet never in the service of power. Deleuze and Parnet illustrate the stakes through the distinction between the trickster's plagiarism, which—albeit critically—operates within fields of similarity and reproduces power structures, and the creative theft of the traitor, the experimenter. It is only the latter who is willing to risk—not only a thought, a friendship, or an affiliation but also becoming unknown, losing both face and identity, and, ultimately, one's life.[7] Yet not every act of treason leads to the loss of a life. Being a traitor can take different forms. There is also the option to kindly refuse, to "prefer not to," to decline operating in the service of power, to deny complicity, and to refuse to stay with the familiar. As we will see, alternatives exist: rhizomatic, errant, synaesethic, embodied, and nonrepresentational thought; theories of indeterminacy; noncontractual relations between wave and particle. These alternatives hold the power to undo the foundations of physics and philosophy alike, delegitimizing not only the metaphors we live by but also the political theories, legal concepts, and institutional narratives by which human and nonhuman beings continue to die. At first glance, it may not seem too dangerous for a thinker, especially an academic one, to stand up to the laws of thinking. Yet Herman Melville's Bartleby, famous for his adamant insistence on "preferring not to," did more than refuse compliance by ceasing to copy legal documents; he went all the way, eventually starving to death. What to do—and how—with all the privilege? Activism, civil disobedience, protest, strike, hacking the system, hijacking the bus, setting everything on fire? Can you pay the price—am I willing to? Everything is a question of power: How far will you go?[8] Deleuze warns that thinking is dangerous. Thought is political and bears risks. For those privileged to inhabit powerful spaces of thought and knowledge production, staying with either the familiar or the trouble is an ethical question. It requires turning against the familiar, the safe, and the hegemonic—the sovereignty of thought, the power of theory, the law of the signifier, and the politics of knowledge. If I turn, will they turn too? Can we turn, twist, and contort together-apart?

Haraway's gesture toward a thought committed to more-than-friendship—more intimate than the love for critique and more allied than the hate of a perceived common enemy—signals the need for an ethics of thought. It advocates for a practice of theory that *matters* not only to our friends—those we recognize most readily and whose lives we deem thinkable and livable—but also to those who bear no similarity to us and to modes of thinking fundamentally different from our own, even unrecognizable to us. Indeed, when Haraway criticizes Latour for using Schmitt to think about climate change and the future of earth, she explains her turn against this intellectual bond with the following: "The question of whom to think-with is immensely material." She cautions: "It matters which thoughts think thoughts. We must think!"[9] Importantly, *matter* here is not to be understood as simply *representational* but rather in its most embodied and most coarsely violated form; namely, the Holocaust committed under the Nazi regime, a regime that was, as we know, upheld not only by actions but also through thought and its thinkers—among them the jurist and law professor Carl Schmitt and the Nazi party official Adolf Eichmann, who, according to himself, only ever thought and never acted, and who, according to Hannah Arendt, *never* thought yet acted fatally.[10] Haraway seeks to understand thoughtlessness precisely in relation to both Arendt's definition in *Eichmann in Jerusalem* and to the Anthropocene as a specific geopolitical juncture that threatens—for different yet related reasons—to reiterate genocide and foster unprecedented ecological devastation. Thoughtlessness is, for Haraway, inextricable from immateriality precisely because "the world does not *matter* in ordinary thoughtlessness."[11] Her reference to Arendt and, by extension, to the convicted Nazi war criminal Eichmann emphasizes the urgency of questioning modes of thinking and sensemaking: "Revolt! Think we must; we must think. Actually think, not like Eichmann the Thoughtless."[12] Importantly, the reference is not simply to thoughtlessness but also to its close ties to language and law and their shared complicity in eradicating both bodies and modes of existence. Indeed, Arendt's *Eichmann in Jerusalem* offers insightful and complicated contemplations on what it means to think (although Arendt stays close to the Kantian understanding) and how thoughtlessness relates to horrendous encounters between stigmatized bodies and blindly (yet consciously and deliberately) applied positive law. It is, as Judith Butler points out, a condemnation of Eichmann's failure to be critical of positive law, his failure to withhold from obedience to a legal system that orders the systematic killing of millions of racialized bodies, diagnosed by Arendt as the "inability to *think*, to think from the standpoint of somebody else."[13] This formulation closely reveals that to speak and think the law means being specifically embodied, having a *standpoint* and at least *some body* that is related and relate-able to other bodies, somewhere else.

Another glimpse into the inextricability of thought, the embodied and material world, law, and the question of response-ability is provided by the written records of Carl Schmitt's interrogation by US lawyer Robert Kempner in 1947. This was not the first time he was arrested, but this time he was accused of being a war criminal and imprisoned for six weeks in a solitary cell in Nuremberg. One of the most interesting accusations was that he had committed an "intellectual crime"; specifically, creating the ideological foundation for wars of aggression, war crimes, and crimes against humanity. Schmitt responded by calling himself an "intellectual adventurer," arguing that "this is how thoughts and knowledge arise." Kempner countered with a question that calls for an ethics of thought that fosters response-ability and resists separating thought from matter: "But what if that which you call a search for knowledge results in the killing of millions of people?"[14] The states of thinking—and, importantly, the dominant understanding of what it means to think—are also pointedly expressed by the Swiss writer Max Frisch. Specifically, his 1958 play *The Arsonists* warns of fascism taking root in everyday environments and thematizes the protagonist's unwillingness to acknowledge the imminent and blatantly visible threat. Everything burns down. Only one question, applicable not only to fascism but also to the refusal of thinking-with what matters and is excluded from mattering, remains: "Knowing full well how inflammable the world is, *what did you think?*"[15]

Thoughtlessness of that kind is not merely an inability to think rationally but also an incapacity to think-with, to entangle, to track lines of living and dying, and to cultivate response-ability.[16] Haraway's "we" in "we must think" and the imperative to "revolt" reach out to her friend Latour, who might have forgotten for a moment that thought actually *matters* and that it has indeed mattered before. It also asks her readers—may they be considered friends or oddkins—to take seriously that thinking is *less* a practice of reflexivity than one of world(s)making (or -destroying), while thoughtlessness—which I will demonstrate to be not the absence of thought but thought that is incapable of being in touch—is a surrender to the conviction that the world, as the subject's Other, does not really *matter*.[17] It is a warning against the romanticization of thought, situating it above the idle matter(s) of existence, bestowing it with the grandeur of transcendence where it can conveniently deny responsibility for what happens on the ground or placing its locus in the figure of the genius, the individual, the thinking subject and exceptional human with which thought itself is said to begin. While I will return to these assumptions later in this section, consider a recent example of how underlying humanistic, and thus exclusive, notions of what it means to think determine what it means to be human and establish (exclusively anthropocentric) normative structures. In the article "Thoughtfulness and the Rule of Law," Jeremy Waldron, a law professor at NYU Law School who teaches legal and political philosophy, argues that "we want to be ruled thoughtfully" and characterizes "thoughtfulness" as "the capacity to reflect and deliberate, to ponder complexity and to confront new and unexpected circumstances with an open mind," and thus as related to human dignity. Waldron's text seeks to parse out "legalistic thoughtfulness" and focuses on three "thoughtful aspects" of the rule of law: standards, procedures, and precedents, which carry with them traditional notions of objectivity. What reveals the underlying assumptions here most clearly, however, is his conclusion, which cites Aristotle: "law is reason unaffected by desire."[18] Waldron professes to be intrigued by the connection between law and reason, which is, he writes, the connection "between law and the god-like activity of reasoning." Whether the concern is thoughtlessness or thoughtfulness, a brief look at how thought is either endowed with normative power or conveniently positioned in a transcendent space, ontologically separated from the embodied and material world, reveals the necessity for an ethics of thought. More concretely, it underscores the need for an ethics of sensing and sensemaking that is attentive to difference rather than sameness, respectful of the unknown rather than reproducing the familiar, and sensitive to the underlying assumptions of every mode of thinking and the matterphorical concepts they bring into the world. As will become even clearer over the course of this book, rethinking what it means to think, and consequently what *theory* can do, will have major consequences for legal thought and concepting too. What if *doing theory*, whether within institutional contexts of knowledge production (such as universities) or outside of institutional contexts, were not seen as a practice of abstraction, logical reasoning, and building concepts with numbers and words but a practice inextricable from the matter, forces, and energies that constitute what we perceive as the world? What if theory, and the modes of thinking involved, cannot be understood in an anthropocentric, perhaps not even a classical humanist, way—and neither can knowledge production? If, as Barad argues, "doing theory requires being open to the world's aliveness," as "theories are living and breathing reconfigurings of the world" and "the world theorizes as well as experiments with itself," how can we let go of the assumptions that underlie Cartesian and representational modes of thinking?[19] The stakes of these questions are profound, primarily because dominant ways of making sense—whether in law, economics, sciences, or the humanities—hold enormous power. They influence not only knowledge production but also the very questions of what can exist and what can be imagined, understood, perceived, and valued—and what cannot. Does thinking, for example, take (a) place and if so, where? Achille Mbembe points in his essay "Decolonizing the University" to the still overwhelmingly dominant presence of the Eurocentric epistemic canon that "attributes truth only to the Western way of knowledge production," "disregards other epistemic traditions," and "tries to portray colonialism as a normal form of social relations between human beings rather than a system of exploitation and oppression."[20] He reminds us that this hegemonic tradition actively represses anything that is thought from its presumed outside. The questions about whether thought can *take a place*, where it can take place, what this place (be it an institution of higher education, an opera, a theatre, a concept, a land, a molecule, the earth, the planet Mars) becomes after a certain mode of thought has appropriated it, and how proprietary thought is in turn reproduced in this place necessarily situate the problem in the context of colonizing practices—including epistemological colonialism. The claimed superiority of this mode of thought (and knowledge production) often engages in spatial exceptionalism, which ontologically positions thought outside of worldly matters where it is believed to eventually transcend all its material constraints: Thought becomes a product of the rational subject and knowledge is recognized only when expressed in certain languages, while futures are written in code for "postbiological" minds that will finally have become intelligent enough to colonize not only *terra* but outer space. This is not simply wordplay but was in the late 1980s declared the core principle of the then-nascent transhumanist thought under the term *extropy*. As one of its founders—back then a graduate student at the University of Southern California—Max T. O'Connor (now known as Max More) writes in the inaugural *Extropy* issue, extropy is the "continual outward expansion of human beings" to seize the "massive quantities of resources waiting to be exploited," inevitably leading to the establishment of space colonies that are "devoted to particular philosophies," ideally "free market anarchy."[21] As More furthermore states, this "off-world revolution" is promoted and anticipated by extropians precisely because of its effects on "intelligence,"[22] expectedly enabling the "transcendent expansion" that will make transhumanists the "vanguard of evolution."[23] It is telling that contemporary transhumanism has been significantly shaped by the aspirations and desires of a group of male graduate students at the University of Southern California, a private university known for its focus on technological and scientific research, and their committed, fraternal friendship. Reading through the *Extropy* issues reveals the friend-enemy binary (the transhuman "us" against those simply not intelligent enough to join the club, as well as "statists" and "deathists") and an inhospitality toward the earth, extending to the notion of a "hostile universe" that needs to be fought and conquered by a new species, transhumans having become posthumans.[24] While the focus of extropist endeavors has always been on creating free market–based and nongovernmental modes of law and governance, their core tenets—reason, intelligence, progress, and radical individuality—are, as transhumanists proudly assert, rooted in European Enlightenment humanism, albeit with a distinct US American anarchocapitalist twist. Such self-declarations should be taken with a grain of salt, but it is important to recognize that many extropians—now better known as transhumanists—have received university training deeply influenced by the economic, political, and epistemological frameworks rooted in Enlightenment humanism. Without disregarding the history of humanist thought all together, the question of what underlying assumptions make its heirs prone to such proprietary, individualist, and representative thought requires utmost attention when it comes to rethinking thought toward more diverse, collaborative, caring, and sustainable ways of existing on a shared, yet not owned, planet.

In addition to the spatial, there is also a temporal dimension of thinking taking (up) p(l)ace. What is the temporality of thinking? How does it construct its epistemic and material histories and how does it relate to others? If, as Deleuze writes, "thought thinks its own history (the past)" precisely in order "to free itself from what it thinks (the present) and be able finally to 'think otherwise' (the future)," then what is the temporality of a thought that allows us to think differently?[25] Certainly, these questions—pertaining to the temporality and spatiality of a thought taking (a) place—are not simply about subscribing to or working against certain schools of thought but require a commitment to challenge a certain kind of academic thought as such. Returning to the predicament of critique and friendship, such an inquiry also demands calling into question what Fred Moten and Stefano Harney call the "critical academic," who may believe they are challenging power relations but who eventually ensures the heteronormativity of academic thought, which is constantly reproduced under the guise of new terms, new specializations, and claims of professionalization.[26] It is both a privilege and a constraint that the critical academic can oppose the university, regimes of knowledge production, and even the very idea of law and constitution without touching the foundations, "without touching one's own condition of possibility."[27] This certainly complicated and at times existential friendship with the institution, expressed in the hesitation that becomes a (academic) lifetime of waiting before the law, and the tacit agreement to the almost mystical source that upholds the conditions of critique have also been pointed out by Peter Goodrich in his "The Critic's Love of the Law." In this essay, Goodrich exposes the paradoxical love and counter-intuitive affection the critical scholar (in England) bears for what they claim to hate and fight the most: the law. The critical scholar's "rapid transition from resistance to law to calls for its renewal" does not only pertain to law but also to the image of thought, the rules of thinking, the state of meaning.[28] If even critical modes of thought (those most intimately linked to power and governance, as Michel Foucault indicates) cannot resist the attraction of the image of thought, what would it take for thought itself, critical or not, to unthink its own constitution? To break that image of thought that relentlessly determines what it means to think?[29]

This question is not new, and given the state of the world, from the catastrophic and cascading effects of global warming to wars, genocide, and ecocide, calls for thinking the future differently have proliferated, including from within Western academic institutions. There might indeed be a greater sensibility for or attentiveness to the long-denied possibility that *thinking* may be too real, perhaps even too impersonal, and too decentralized to be contained by the Enlightenment concept of reason (or its interpretation) after all. Thinking might also be too anarchistic to be kept in check by institutional hierarchies, too entangled with difference to be captured by analogy, and too committed to embodied relationality to be mobilized for out-of-touch formulas, laws, political systems, and economic models.

But how can we—as indeterminate and fluid space, constantly shifting and reconfiguring our relationality—think not only the future but also thought itself differently? How can we create the conditions for thought to inhabit relationality, operating unconstrained by binary dualisms and dialectical negations, creatively modulating complicity, and making possible different modes of existence? How can thought break free of its constraints, being neither *of* the subject nor *about* an object, and become nomadic thought, or what Deleuze and Guattari describe as a "vegetal model of thought," Édouard Glissant calls "errantry thought," and Fred Moten and Stefano Harney term the "undercommons?" How can we sense and make sense of the material and embodied accountability, let alone response-ability, that accompanies each act of creative treason—each gesture of trust in what is unfamiliar and unknown to the sensorium of contemporary power structures? How does an ethics of sensing and making sense—*synaesethics*—not only differently but through and with difference commit to the ongoing making (and unmaking) of not just ideas and conceptual frameworks but modes of existence: ways of living and dying, becoming and perishing, breathing and suffocating, entangling and disentangling, down to the very indeterminacy of thought and being? The neologism *syn-aesethics* suggests that ethics begins with a reconceptualization of thinking itself. Rather than being an individualized practice centralized in the mind of a subject—akin to the philosopher whose thoughts need no air, as criticized by Luce Irigaray—thinking is seen as a collaborative and decentralized mode of sensing and sensemaking.[30] The prefix *syn-* signals the relational aspect expressed by *thinking-with*, where both the hyphen and *with* denote a multidimensional relationality that extends beyond a simple linear process. The prefix *syn-* and the verb *thinking-with* indicate that the nature and operation of thinking are shaped by the particularity of collaborations and encounters (relations, differences, intensities, sensings). Thinking-with certain indigenous modes of thought, which understand knowledge as relational and shared with the whole cosmos, creates different worlds than thinking-with Cartesian modes. Thinking-with feminist thinkers who refuse to reproduce patriarchal patterns enables different modes of thought than thinking-with Schmitt or Martin Heidegger. Thinking-with critical posthumanists allows for different relations to space and place than thinking-with transhumanist ideology. Thinking-with markets, organisms, weather patterns, histories, forces—each mode potentially alters modes of existence on the ground. Speaking not only of an ethics but of a synaes-ethics draws attention to the fact that any form of making sense—that is, of engaging in meaning-making practices—requires attentiveness to the ongoing intra-action of modes of sensing and the being of the sensible (i.e., that which is sensible but not recognizable). Thinking, far from being restricted to human and representative thought, (un)matters and (un)makes-sense. As such, it is both an onto-epistemological and ethical concern. The dynamic field, or, as Deleuze and Guattari call it, the plane of immanence, that becomes accessible—sensible and sense-able—is one in which matter and meaning, body and mind, are inseparable. *Synaesethical* engagement with this field, in concept creation, analysis, design, prototyping, and other processes, is what I call *matterphorics*.

As will become clearer through the next chapters, matterphorics as a mode of doing theory and creating concepts is crucial for law and legal thought—which is still deeply embedded in social contract logic, guided by centralized power structures, productive of abstract concepts, and reliant on representational modes of thought. These modes assume that thought exists—if it is accepted as existing at all—in a separate human and sometimes divine realm of consciousness, untouched by physical forces, molecular bonds, and other matters of real alliances. Yet such thought, assumed to be transcendent, can determine who and what matters. It is precisely these determinations of meaning—performed from a presumed outside and excluding not only matter itself but also other possible meanings coming to matter—that constitute acts of capture and appropriation.[31] Within the field of law, rethinking what it means to think also alters the conditions of both the forms of violence and the creative potential inherent in legal thought, modes of governance, and established as well as emerging normativities. What, we might ask, remains of the modernist notion of law, including international law, if analogy and comparison lose their power? How does law operate in an immanently entangled world, sensitive to gravity and in alliance with oxygen molecules instead of gaining its force(s) and power(s) from representational thought, from origins created and guarded by sovereigns assumed to be above both the law and the world? If the right to life is no longer solely linked to the capacity to reflect and to express thoughts in certain languages (or in language in general), then how can law relate to life and modes of existences that express themselves nonlinguistically? If established modes of thought and knowledge production, including critique, evolved with centralized forms of power, do not decentralized forms of power and governance necessarily demand different modes of analysis and thought?[32]

The claim that carries the argument of this part of the book, and the legal questions resulting from it, is that by not obeying a false sense of loyalty to philosophy's *philia*—which seems, at least in its European tradition (and its respective variations), reluctant to give up its exclusiveness—different modes of existence might become thinkable and live-able by committing to the *matter(s) of thought* and, consequently, to a mode of theory that matters. In the end, we might discover that those most devotedly pledging allegiance to a mode of thought inextricable from notions of progress, growth, individualism, and reason are those departing the most from an ethics geared toward livable thoughts and thinkable lives.

# 2

# Metaphor and Nuclear Equations

## ON THE SUNNY SIDE OF THEORY

$E = mc^2$
—ALBERT EINSTEIN

The very emergence of justice and law . . . implies a performative force, which is always an interpretative force. —JACQUES DERRIDA, "Force of Law"

The liar is a person who uses the valid designations, the words, in order to make something which is unreal appear to be real. —FRIEDRICH NIETZSCHE, "On Truth and Lies in a Nonmoral Sense"

*Staying with the Trouble* is not a book of representational thought but of learning to "make kin in lines of inventive connection as a practice of learning to live and die well with each other in a thick present."[1] Consequently, it does not have a fixed object or offer the perspective of the detached observer and judge, analyzing a world pre-made. In Anna Tsing's book *The Mushroom at the End of the World*, pigeons, fluids, and string figures move and mingle, reaching out to a myriad of wandering and spreading mushrooms that too refuse to be arrested in thought and analyzed to death. Both books are furthermore oddkin to Renisa Mawani's *Oceans of Law*, which, instead of reiterating familiar narratives based on common territorial concepts, asks thought to become-ship, so that the ship, the *Komagata Maru*, and its movement can tell the colonial legal history that both maps and archives have rendered unspeakable. The currents to which the book draws the reader's attention not only move decolonial thought (and the ship) across the ocean but also "connect the ocean regions that have long been divided in European thought."[2] Importantly, neither the currents nor the ocean nor the ship are simply metaphors; they draw on "materiality to trace the legal overlaps between ocean arenas and the movements of colonial law."[3] All three books counter—to different degrees, with different means of form and style, and in terms of what they engage with—representational and anthropocentric modes of thought, seeking to give voice not to objects but to the very relationality of mind and body, word and world, matter and meaning. Giving voice, however, must not be understood as transfer of rights or ownership, let alone a noble gesture toward the Other, but as expression of what Andreas Philippopoulos-Mihalopoulos terms "spatial justice." Serving as "a deep-seated desire to eradicate . . . the desires that have placed us central and made us vacuous effigies of human preponderance," spatial justice emerges as a withdrawing. Withdrawing, as Philippopoulos-Mihalopoulos reminds us, is not a passive activity and "has nothing to do with removing oneself from the conflict or sacrificing oneself to a moral priority of the Other."[4] Rather it is a "moving away" from the "desire to carry on with the comfort offered by supposedly free choices, power structures or even by fate," which can only ever be a moving "deeper inside," a giving-*in* to immanence and to being-with rather than thinking-about.[5] Such modes of moving synaesethically, sensitive to the relationality and inseparability of matter and meaning and in search of "another justice, another movement, another space-time," are precisely what matterphorics seeks to emphasize.[6] Before looking more closely into matterphorics, I wish to highlight the traps and complicities of representational thought and its allies: metaphor and analogy. The latter two are certainly powerful bridging devices, mediating meaning across the thereby accepted and reinforced abyss between matter and mind. Yet it is this acceptance that continues to control who or what can come to matter (or is excluded from mattering). Modes of thought—and theory—do not fall into the categories of right/wrong. What matters is what a mode of thought can *do* and where it *takes* its power from. And what could possibly be more apt than the sun and nuclear fission to shed light on the power(s) of thought and sensemaking?

Power is enticing, physically attractive.[7] It charges, forces, draws. Even the best intentions can, at times, wane in the face of power and the heat of the argument. There are certainly more powerful influences than philosophers and academics. Indeed, at a time when the effects of global warming are significantly shifting social and material conditions for existing on planet earth, the aftermath of the last global pandemic continues to materialize, inequality and wars are on the rise, and powerful, novel modes of governance are facilitated by new digital technologies, it might seem trivial to focus on the question of what it means to think and to make-sense. However, we must not underestimate the consequences of how modes of thought operate, the tricks they play when faced with crises or in search of power, and how their assumptions express a deep onto-epistemological normativity: determining what matters and what does not.

It was not long ago—1984 to be precise—that the influential poststructuralist philosopher Jacques Derrida was invited to speak at a conference on "nuclear criticism," which was meant to provide the concerned critical thinker with a rhetorical toolbox to face, and ideally also prevent, the nuclear apocalypse—that is, the absolute destruction of (human) life. In his paper "No Apocalypse, Not Now (Full Speed Ahead, Seven Missiles, Seven Missives)," Derrida—blind to the actual ongoingness of nuclear war in places out of his sight and thought and faced with a natural force that might actually deconstruct atoms—sets out to bestow a group of humanities scholars with a rhetorical power on par with that of nuclear force.[8] The sense of powerlessness (of the critical scholar and academic intellectual) in light of nuclear destruction is compensated by an attempt to transpose matters of matter into those of language, representation, and metaphor, which in turn enables the invention and mobilization of rhetorical weaponry (for example, "weapons of irony"). Derrida's playful description of nuclear weaponry as "fabulously textual," a "fabulous specularization" that only exists in thought and can therefore be conveniently challenged by thought expressed in language, is possible only because Derrida assumes a nuclear war that has not yet taken place.[9] Caught within his rhetorical manoeuvre, he sees no issue in calling upon humanities scholars' allegedly neglected "responsibility" to concern themselves with the "nuclear issue" precisely because, he writes, "we are representatives of humanity and of the incompetent humanities" and because "the stakes of the nuclear question are those of humanity, of the humanities."[10] The assumption that the ontological and epistemological understanding of *human* in *humanity* and in the *humanities* is the same, allowing for the comparison and even equalization of humanity (as mortal human beings in their collectivity) with the humanities (a particular geographically and politically constructed epistemological field), lays bare the bias of representationalism. The consequences of Derrida's conscious move to mobilize representational (in particular, metaphorical) thought become legible when we look at what he declares as most threatened by the anticipated nuclear catastrophe: the absolute destruction not of bodies of all kinds but of *literature* as "the body of texts whose existence, possibility, and significance are the most radically threatened, for the first and last time, by the nuclear catastrophe"— and with it the self-referential logic of representationalism, the archive, the name, and the text. In other words, it is the destruction of the symbolic order and a particular condition of representation that the humanities scholar is, and in fact *can* be, most concerned with.[11] More than nuclear *criticism*, Derrida's text reinforces a mode of nuclear *equation* that he had already justified in an earlier text, "White Mythology: Metaphor in the Text of Philosophy," and written about ten years before his delivery of the "No Apocalypse" paper. To demonstrate what is meant by *nuclear equation*, I wish to point to a specific moment in the text when Derrida reads Aristotle's account of metaphor in his *Poetics*.[12] Aristotle defines metaphor as that which "consists in giving the thing a name that belongs to something else"; that is, a "transference . . . on the grounds of analogy."[13] For Aristotle, analogy is metaphor *par excellence*, yet he grants that analogy can at times be inventive. As an example, he mentions the absence of a term that denotes the sun's generating force: "to cast forth seed-corn is called sowing; but to cast forth its flame, as said of the sun, has no special name."[14] The analogy between seed and sun, both considered life generating, leads to the poet's phrase "sowing a god-created flame."[15] Derrida's aim is to show that there can be no invention as every word and metaphor already draws upon another. However, Derrida, thinking-with Heidegger, uses this example to refute the possibility of an exteriority to metaphor and to claim that arguing otherwise would concede to the logic of appropriation, land taking, and foundation building. Even the sun itself, he continues, is always already metaphorical: "With every metaphor, there is no doubt somewhere a sun; but each time that there is the sun, metaphor has begun. If the sun is already and always metaphorical, it is not completely natural. It is already and always a lustre: one might call it an artificial construction if this could have any meaning in the absence of nature. For if the sun is not entirely natural, what can remain in nature that is natural?"[16]

The essay undoubtedly fights many battles. Among them, it seeks to demonstrate the inseparability of philosophical concepts and metaphors, to claim the impossibility of direct access to things that are mediated by discourse and its tropes, to counter naturalized meanings, and to refute any exteriority of either metaphor or language, which makes it possible for him to state that even "theory is a metaphor."[17] Despite the critique of proprietary thought, it is crucial to imagine further implications of Derrida's metaphorical thought and the representational power it unleashes, both of which are reinforced in his 1989 Cardozo Law School talk on the force of law. When theorizing about the "metaphor of metaphors" or the "mystical foundation" of law, Derrida seeks to expose its functions and, to speak in Spinozian terms, to reveal its power (what it can *do*).[18] Derrida's deconstruction does not question the assumption of an immaterial, groundless foundation, the first cut that positions word over world, as such. Such critique would bereave deconstruction of its power. Nor does he simply deconstruct the concept of law. In a strategic rhetorical move, Derrida equates deconstruction with justice and depicts difference as "difference of force" and "force as *différance*."[19] This deferral of meaning as a construction of semantic undecidability or indeterminacy—in any case, a maneuver that takes place through the representational layer of language—serves as a rhetorical force, a power move, as well as a preemptive limitation of response-ability. What lies beyond representation? How do we navigate force fields that undermine the condition of represent-ability? How do we become response-able to forces that perform existential exclusion without any representational traces?

In "White Mythology," Derrida refers to Friedrich Nietzsche's claim that truth is nothing but a "mobile army of metaphors, metonyms, and anthropomorphisms"; that is to say, truths are "illusions about which one has forgotten that this is what they are; metaphors which are worn out and without sensuous power."[20] Derrida does so less to reject the power of metaphor and language—indeed, in "No Apocalypse," he invokes this power—than to indicate an economy of exchange and the production of surplus value that he further traces in Ferdinand De Saussure's work. As mentioned, Derrida's inquiry is guided by his attempt to demonstrate the inescapability of metaphor (and consequently also of language and text) and to deny the existence of a real and external referent. He aims to show that even "an object which is the most natural, the most universal, the most real, the most clear, a referent which is apparently the most external, the sun," always participates in a "process of axiological and semantic exchange" and therefore does not "escape the general law of metaphorical value."[21] My point here does not so much pertain to Derrida's deconstruction of the sun as metaphor for Western modes of knowledge and concept production or to challenging his theory of metaphor itself. Rather, I wish to point to the consequences of metaphorizing the sun and its physical forces, which takes its cue not only from the predicament of naming and denoting but perhaps even more so from an incapability of confronting the insufficiency of human knowledge and agency when it comes to physical, nonhuman forces. Rather than deconstructing white, Eurocentric power, so to speak, Derrida's essay sets out to construct and legitimize the source of a power that, albeit rooted in representation, claims its force from the sun—the very *same* sun—and handily comes to aid in 1984 when Derrida seeks rhetorical means to counter the threat and presence of nuclear force. This is precisely what allows him to oppose nuclear weapons to "weaponry of irony," or to send missives as missiles.[22]

Theorists, too, choose their complicities. Whether considered casualties in the struggle for representational justice or acknowledged as symptomatic in representational and language-based modes of thoughts, such "fabulous speculations" require a Cartesian consensus pointedly exposed by Fred Moten and Stefano Harney.[23] "Never having to confront the foundation, never having to confront antifoundation out of faith in the unconfrontable foundation," the critical intellectual "can float in the middle range" or, we might add, inhabit the realm of mediation.[24] As I will demonstrate, it is precisely the consensus on the unquestioned foundation (the image of thought) and the thereby upheld reign of representationalism that deprive alternative, nonrepresentational modes of thought of not only a ground but of air, mixtures of gases, the alchemy of photosynthetic breath, silent dances of cacophonous multiplicities, wave particles, speaking electrons, and sounding senses.

While approaching "theory" as the deliberate mobilization of language and representational thought (which was certainly also owed to the linguistic turn) was a popular approach at the time of the Cardozo conference, alternatives existed even within the realm of critical scholarship. Indeed, a conversation between Gilles Deleuze and Michel Foucault titled "Intellectuals and Power" and recorded in 1972 (two years prior to the publication of Derrida's "Mythology" essay and twelve years prior to the conference) calls for a different understanding of theory. The intellectual and the theorist, Deleuze argues, have ceased to represent or be representatives while those involved in political struggle have ceased to be represented. "There is no more representation. There is only action, the action of theory, the action of praxis, in the relations of relays and networks."[25] For both Deleuze and Foucault, theory is inextricable from power. Theory, Foucault argues, is local, regional, and nontotalizing. It is "a struggle against power, a struggle to bring power to light," a "struggle to undermine and take power side by side with those who are fighting, and not off to the side trying to enlighten them."[26] Deleuze adds that theory therefore "has nothing to do with the signifier"; it must be used, it has to work, and it multiplies. Indeed, "it's rather in the nature of power to totalize," while theory, Deleuze states in agreement with Foucault, "is by nature opposed to power."[27]

Derrida is not the culprit, but he is complicit in using the power of representationalism, and he is not alone. It is, in fact, not unusual for metaphor to make an appearance when physical forces seem uncontrollable, when the spirits invoked refuse to obey human commands and logic. When recalling "Trinity," the nuclear test—and first detonation of an atomic bomb—in New Mexico on July 16, 1945, J. Robert Oppenheimer notoriously stated that "if the radiance of a thousand suns were to burst at once into the sky," it would be "*like* the splendor"—yet, as we will see, not *like* the metaphorical "lustre" in the previous Derrida quote—released by the nuclear bomb's detonation.[28] Oppenheimer, despite his fatal hubris, was aware that only metaphor can approximate the physical phenomenon of a force so utterly incomprehensible to human intellect and imagination, but he also knew—and carried (out) this knowledge with utmost pride—that there is another relation between the forces of the sun and nuclear weapons, one that is neither metaphorical nor metamorphizing but lethal. We don't know how much Aristotle knew about the actual physics behind the life-generating forces of the sun. But we can be certain that Oppenheimer knew that nuclear fission, the process that causes the release of energy responsible for the destructive force of the first generation of atomic bombs—including those which were detonated over Hiroshima and Nagasaki—generates heat of one hundred million degrees Celsius, a temperature assumed to exist at the center of the sun. By the time the nuclear criticism conference took place, it was also known that the sun, as well as all stars, generates its energy through nuclear fusion, which is the main process underlying the hydrogen bomb. In fact, the hydrogen bomb tested by the United States at Bikini Atoll in 1954 was over one thousand times more powerful than the nuclear bomb US air forces dropped on Hiroshima on August 6, 1945. Unlike fission bombs, there is also theoretically no limit to the force of a hydrogen bomb. What these examples show is that Barad's early claim that "the power of language has been substantial," even "too substantializing," is a crucial one and exceeds the oppositional framework (matter *or* word/language) in which it is commonly understood and discussed.[29] Indeed, in "After the End of the World: Entangled Nuclear Colonialisms, Matters of Force, and the Material Force of Justice," Barad shows the entanglement of theory and the all-too-real material force of the twenty-three nuclear bombs dropped on Bikini Atoll. Theory, they argue, must neither be disentangled from matter nor from the potential violence it entails: "The theory and the bomb inhabit and help constitute each other."[30] This is not only true for quantum theory. Modes of thought are entangled with force(s), emergent from fields of power and potential, and thus are always also complicit with histories of ineffable violence as well as creative resistance.

Ontological separability—of mind/matter, signifier/signified, non/human, particle/wave—solidifies the foundation of representational thought, the thought that precedes and determines who and what can come into existence or is excluded from it: sovereign thought before the law and after the fact of its own constitution. It is thus crucial, as Barad argues, to understand that physical forces, including nuclear forces, are not separate from what is referred to as the realm of the political—not least because they call into question the "foundational assumptions" of representationalism and "separability itself."[31] Matters of thought and theory can be neither reduced to mere reflection nor confined to the imagination of the thinking subject: "The foundation can never resemble what it founds" and "it does not suffice to say of the foundation that it is another matter."[32] To respond to Derrida then, we might concede that even if theory is understood as a metaphor, what is mistaken for a metaphor (that is, for a rhetorical device) ought too to be understood *matterphorically*: as a nonrepresentational, onto-epistemological expression. In other words, theory and thought must be considered nonrepresentationally, in acceptance of that fact that, as we will see, an electron *can* crash into a language.

There is no doubt that metaphors are powerful in the way they create worlds of meaning—as if the *world* were only an image or a reflection, as if thought in the end wouldn't actually *matter*, as if difference must be mediated to be thinkable,[33] as if the question about what thought is wasn't always already linked with an "and" (*and* matter, *and* molecule, *and* physical forces, *and* power, *and* life . . .). Indeed, as if, as Edmund Husserl claimed to Deleuze's consternation, the tree can burn down and dissolve into its chemical constituents while its meaning and sense, owing to their lack of chemical constituents, forces, and real attributes, cannot.[34] That is also to say: As if language wouldn't also crash into electrons, reacting with and forming new molecules, elemental bonds that will have mattered differently. The imperialism of language in its function as "universal translator and interpreter," as denoted by Deleuze and Guattari, explains why Derrida fears mostly not the end of bodies, not their combustion and evaporation, but the end of representation or, even more precisely, of the signifier, for the impossibility of its going up in flames is the foundation of the meaning and sense of the world.[35] Meaning, in this representationalist view, is claimed to matter apart from the matter(s) at stake—and this despite the fact that it is precisely matter that, following quantum field theory, is at stake. Indeed, matter, Barad reminds us, not only became mortal but was exploded and murdered with the atomic bomb when the "smallest of smallest bits, the heart of an atom, was broken apart with a violence that made the earth and the gods quake."[36] It is, in the end, so to speak, the "imperialism of the signifier" that not only enforces the imperialism of language but that is in fact "affecting all regimes of signs," reducing expression to the signifier.[37] This is also why, in their collaborative work, Deleuze and Guattari set out to do away with metaphor—and with it, all symbolism, all signification, and all designation—in favor of metamorphosis. Starting from the middle, their chosen point of entry is Franz Kafka's work, which "kills all metaphor" so that there is "no longer any proper sense or figurative sense, but only a distribution of states that is part of the range of the word," and no longer things, but intensities overrun by deterritorialization.[38] Thinking nonrepresentationally, expressing despite language, and reading signs nonimperialistically demands what Deleuze and Guattari call the "abolition of all metaphor" and the acceptance that "all that consists is Real."[39] While there is no consensus in matterphorics and no rulebook guiding synaesethics, both operate on this principle.

As of this day, there has not been much in-depth research on the effects of radiation and nuclear contamination on marine life. There is no study on what radiation might *mean* to and for a brittle star. Unfortunately, I cannot offer scientific insights that could address this absence. It is beyond my capacity to imagine what a metaphor might be or mean for a brittle star. It is known, however, that *Ophiomastix wendtii*, a specific brittle star species, is very light sensitive and hides from sunlight. What is also known is that it has no brain, and yet thinks. One might wonder if it stands to reason, so to speak, that we should articulate a theory on how to think and encounter language differently *with* the brittle star. Barad, in their essay on the ontology of knowing titled "Invertebrate Visions, Diffractions of the Brittlestar,"[40] makes a case for why a theory that denies metaphorizing, representationalism, and the violence of rendering the material world unknowledgeable, thoughtless, and incapable of theorizing is precisely what would stand to, perhaps stand *up* to, reason. For one thing, as mentioned, the brittle star has no brain; knowing and being, thinking and living, are entangled. It does not suffer the Cartesian doubts of an alleged mind-body split.[41] This means that a notion of theory that relies on the superiority of the thinking, transcendent mind or claims its legitimation in an alleged cognitive superiority of the human brain cannot find a hold here, at the sandy bottom of the sea. The essay is based on scientific findings on *Ophiomastix wendtii* according to which the brittle star's skeletal system is composed on an array of microlenses on its surface that constantly collect and focus light on points that correspond to the brittle star's nerve bundles, constituting a system similar to a compound eye. That is to say, the brittle star is "a living, breathing, metamorphosing optical system" for which "being and knowing, materiality and intelligibility, substance and form entail one another." It does not "think much of epistemological lenses or the geometrical optics of reflection." It thinks diffractively.[42] Given that, in its etymology, theory denotes "seeing" or "viewing," thinking-with the brittle star *makes* sense. If it engages in a mode of theory based on diffraction rather than reflection and reflexivity, the optical phenomena guiding representational modes of thought and knowledge production, it is rather unlikely that it will aim to enlighten a subject or cast a spotlight on the human (or itself) while leaving the rest in the dark.[43] In fact, as Barad argues, the brittle star troubles the subject/object and self/other distinction with its very mode of being. Its differential materializations (e.g., changing its color in response to light, breaking off its body parts when in danger and regrowing them) are material-discursive—that is, they are boundary-drawing practices that constantly renegotiate its relations to its environment.[44] For the brittle star, Barad shows, thinking and knowing are not mediated activities: "Knowing is a direct material engagement . . . The entangled practices of knowing and being are material practices. The world is not merely an idea that exists in the human mind. To the contrary, 'mind' is a specific material configuration of the world, not necessarily coincident with a brain. Brain cells are not the only ones that hold memories, respond to stimuli, or think thoughts."[45]

It is particularly the latter argument—that it is not only brain cells that think thoughts—that prompts the question guiding the next chapter: What actually is that which thinks thoughts? How can we understand this thinking of thoughts that lies in between being and knowing, ties them together, is enabled by them, and yet remains distinct? Or, put differently, what would it mean to articulate a theory of thought as a practice of thinking that must remain in touch with the world—of which it is a part but which it does not contain or exhaust—rather than striving to transcend it? How can such a theory take seriously its commitment to not only creating livable futures, pasts, and presents but also sustaining a present that can be inhabited by more than what we already know and deem knowledgeable? Finally, perhaps more out of habit than necessity, what exactly is reason?

# 3

# Matterphorics of ?-Human(ism)

## THEORY FOR LOYAL TRAITORS

All things with conceptual dimension are like language, as all grey things are like elephants.
—ROBIN EVANS, "Translations from Drawings to Buildings"

Reflection on what it is like to be a bat seems to lead us, therefore, to the conclusion that there are facts that do not consist in the truth of propositions expressible in a human language. —THOMAS NAGEL, "What Is It Like to Be a Bat"

And I'm like an earthquake / When I get thinking too much / Everything's falling and shaking and falling / Through my hands / Well maybe I should change my point of view. —PLANTS AND ANIMALS, "Lightshow"

Like *Staying with the Trouble*, *The Mushroom at the End of the World*, and *Across Oceans of Law*, Barad's brittle-star essay is, as might already be abundantly clear, not one of representational thought. However, it has been attributed to post-humanist thought. Although the practice of thought and reasoning that it presents is certainly far from a canonical understanding, it of course resonates with other concepts that share its unconventionality. Indeed, multiple potential histories of different modes of thought are lurking in the folds and gaps of the hegemonic ideas of the thinking mind: the disembodied, transcended subject; the conscious, rational, and sentient human being equipped with the exceptional faculty of logical reasoning that allows for rational decision-making; and the dominance of language as the main indicator for the capacity of thought and intelligibility. Thinking and knowledge production are collaborative and relational practices and processes. Given the dominant structures of representational thought, the epistemological laws it guards, and the ontological assumptions it reflects, thinking nonrepresentationally, expressing different modes of existence in their relationality, and sensing matter(s) meaningfully require unexpected encounters, collaborations, and modes of being in touch. Before looking (or sensing) closer into the matterphorical expressions that open up possibilities for a different mode of thought—anarchic reason (Alfred North Whitehead), the "plane of immanence" (Deleuze and Guattari), the void (Barad), and "Relation" (Édouard Glissant)—I wish to, at least for a moment, address *posthumanism* as mode of thought, often also called a theory. I consider this crucial for two reasons: The first pertains to the abundance of circulating terms—such as posthuman, posthumanist, transhuman, anti-human, nonhuman, more-than-human—and the conflation that often results from it. The differences are not only significant; they have their material and discursive (and material-discursive) histories and canons. Again, the thoughts with which we think thoughts matter. A theory that builds on ecofeminism and a concern for embodied knowledges differs from one based on colonial practices and the notion of the earth as passive surface in ways that exceed terminology. The human in post*human* and that in trans*human* are neither the same nor comparable. Similarly, the temporality and positionality of *trans-* and *post-*, as possible prefixes to not *human* but *humanism*, are different. The relations between the human, humanity, and the humanities are far more complex and multiple than the terminological proximity suggests. However, a theory informed by Western feminisms is of course also different from one that comes out of indigenous thought, just as American Black feminism is different from that rooted in Indian seed justice movements. And of course, each of these movements differs internally. This speaks to the second reason: the need to situate posthumanism in its geographical, epistemological, and ontological context in order to understand its aims. While terms denoting theoretical tendencies and concepts are always internally heterogeneous, they also share ethical, methodological, and epistemological characteristics. Rosi Braidotti, for example, speaks of a "posthuman predicament" as the convergence of posthumanism and postanthropocentrism requires a critique of both "the humanist ideal of 'Man' as the allegedly universal measure of all things" and of "species hierarchy and human exceptionalism."[1] As such, posthumanism—or what Braidotti calls *critical posthumanism*—builds on radical epistemologies such as (eco)feminism, queer and gender studies, and postcolonial studies. Yet it also attends to the ontological relationality that has been given little attention in social-constructivist theoretical accounts.[2] The necessity of correcting this inattentiveness follows from the precarity of earthly modes of existence as a product of capitalist modes of exploitation, fast technological developments, the idea of the Anthropocene as a geological epoch that necessarily denotes uncontrollable natural forces unleashed by human action upon the earth, the undeniability of the material inseparability of human and nonhuman bodies, and the crumbling of the Eurocentric notion of an assumed separability between epistemology and ontology.[3] As Barad writes, posthumanism does not "presume the separateness of any-'thing,' let alone the alleged spatial, ontological, and epistemological distinction that sets humans apart," but can be understood as a "thoroughgoing critical naturalism" that recognizes "humans as part of nature and practices of knowing as natural processes of engagement with and as part of the world."[4] In short, we can understand posthumanism as a commitment to an affirmative ethics of onto-epistemological relationality that necessarily departs from traditional ideas of humanism and anthropocentricism and the economical, legal, and political systems built upon them.[5]

The question of (onto-)epistemological relationality also speaks to the necessity of situating posthumanism. As a mode of thought, posthumanism urges us to rethink our academic loyalties and royalties and build collaborations across temporal, geographical, epistemological, disciplinary, and institutional borders. The minor canon of contemporary posthumanist thought is mainly informed by process-philosophical approaches found in the work of, for example, Baruch Spinoza, Alfred North Whitehead, Henri Bergson, Gilbert Simondon, Gilles Deleuze, and Félix Guattari. From a contemporary standpoint, a list of male European philosophers does not immediately seem to be a radical choice to rethink thought. Yet in their own respective times and contexts—and even in their absence from most of the philosophy curricula taught at universities in Europe and North America—these works and the modes of thought they provide are, in Deleuze's words, traitors to the world of dominant significations and the established order.[6] In this sense, they are *already* and *again* post*humanist* lines of thought, urging us to read histories of thought, knowledge, and life differently and from the perspective of difference. It is, without a doubt, difficult to be a traitor because, as Deleuze writes, it is to create and to experiment. It is not to become known but to "lose one's identity, one's face, in it . . . to disappear, to become unknown."[7] However, one is never *only* a traitor. Rather, one is a traitor *to*: *to* a specific order, *to* a specific idea, and thus in relation to what is being betrayed. Critical posthumanism engages precisely in this traitorism because, as Simone Bignall writes, posthumanism is "a form of self-correction that originates predominately within Western philosophies," challenging their anthropocentric and humanistic approaches to thinking and knowledge production.[8] It is therefore crucial to keep in mind that the *post-* in critical posthumanism is to be taken as an indicator of a specific situatedness in the Western epistemological and philosophical canon. Indigenous epistemologies and ontologies, among others, do not require the *post-*. In fact, they must not be made subject to the humanist narrative and its specific temporality. As Braidotti and Bignall argue, the posthumanist and ecofeminist efforts to "describe a mode of thought adequate to the complex material energies of the earth" ought to be seen as existing alongside a far older indigenous philosophy that "understands the power and potentiality of thought as being materially embedded in the geoformations and trans-species influences that shape and define existence in relational terms."[9] Humanist ideas of knowledge and thinking are, in fact, quite narrow. As Shawn Wilson writes, Western notions of knowledge as an individual entity that can be gained and even owned by an individual researcher are fundamentally at odds with most indigenous conceptions of thought, knowledge, and research. Indeed, in indigenous epistemologies and ontologies, knowledge is conceived as *relational*: "Knowledge is shared with all of creation. It is not just interpersonal relationships, not just with the research subjects I may be working with, but it is a relationship with all of creation. It is with the cosmos, it is with the animals, with the plants, with the earth that we share this knowledge. It goes beyond the idea of individual knowledge to the concept of relational knowledge."[10]

Seen from a nonhumanist perspective then, Barad's onto-epistemological argument that all life forms do theory and that we have to find ways to be in touch and do collaborative research seems less unconventional. In fact, the concept of thinking-with requires precisely this relationality and the collaborations that consequently unfold from it.[11] This also means that situatedness is not just an epistemological concern. It is crucial to acknowledge that concepts have to come-*from*; that is, they are not severed from their histories and specific entanglements. It is also crucial to note how unlikely it is that concepts considered meaningful within academic knowledge production come-*from* precisely the place in which they need to come-*from*; namely, the most precarious worlds.[12]

This is all to say that the (onto-)epistemological relationality that guides post-humanism seeks different intellectual collaborations within the Western canons and acknowledges the necessity and potential of thinking-with less recognizable or even utterly unfamiliar modes of thought and existences. Becoming unknown is a commitment to challenge humanist assumptions that determine what thinking is, who it is that can think, and, based on that, who or what is rendered (un)thinkable. For Barad, critical posthumanism therefore does not start from an already given concept of the "human" but considers the alleged cut between human and nonhuman "a constitutive part of what the theorizing or the analysis entails."[13] Perhaps the more appropriate term, signaling the need for ongoing questioning and acknowledging the histories and critical voices of Western humanist thought, would be *?-human(ism)*. Old and new collaborations unfold problems anew. For example, in *The Order of Things*, in reference to Nietzsche, Foucault notes that "it is no longer possible to think in our day other than in the void left by man's disappearance," for "*this void does not create a deficiency; it does not constitute a lacuna that must be filled in.*" It is, he concludes, "nothing more and nothing less than the unfolding of a space in which it is once more possible to think."[14] In understanding my approach as emerging from a ?-humanist concern about modes and matters of sensemaking, I wish to attend precisely to the unfolding of a space (or spacetime) that makes it, once more, possible to think. This requires a sensitivity to the dynamism underlying the prefixal use of ?, which in turn calls for the ongoing questioning of the ontological and epistemological assumptions that guide dominant modes of thinking.

Making-Sense-With the Brittle Star:
Matterphorical Expressions

Without doubt, Derrida was convinced that language matters. In fact, his work is devoted to the significance of language. Ironically, that significance is precisely what makes deconstruction insensitive to what *matters*, as it neither asks what the processes of materialization (in and of the world) are nor considers the very possibility of the signifier *really* touching what it claims to signify. According to that notion of significance, which relies on signification and is inattentive to the ontological assumptions that it thereby accepts, an electron *cannot* crash into language. The question makes no sense. But what if it could?

That this is still an all-too-sensitive topic within various academic circles can be grasped from the many disputes about whether deconstruction is a "textual practice" or goes beyond the text.[15] Friendship, once again, marks its territory.

In a printed conversation with Derrida titled "Following Theory," the literature scholar Christopher Norris deplores a "notorious sort of simplistic, naive misreading" of Derrida's "il n'y a pas de hors-texte" claim according to which Derrida is "some kind of transcendental idealist who doesn't believe that there's any reality beyond language or the play of textual representations." Objection to his judgment is precluded in advance by Norris himself: "I think we can discount those vulgar misreadings, especially after what we heard this morning."[16] I was not present that morning. In any case, my question is not concerned with whether deconstruction goes "beyond the text," for it already misses the point by not questioning the assumptions underlying both deconstruction and the mode of representational thought it builds upon. But I keep wondering what it might mean for modes of theorizing and thinking that make-sense of (and ideally *with*) language if an electron were able to crash into (a) language.

I consider these concerns central to synaesethics: not only *whether* or *not* language matters but also *how* and for *whom*, or *what* has been contested, fought over, stated, and restated. The stakes are high. Far from being theoretical, these questions pertain intimately to thought. Language and thought are historically interwoven, at times so closely that those rendered unable to speak the right language, or to speak it correctly, were said to be, by nature, unable to think and thus prevented from breathing. Hanging from trees, lying on the ground, being blown up, sinking into deep waters: *Cogito, ergo sum*, written in Latin, translated into French, English, German, Spanish, Portuguese, excluding the thoughtless as well as the unthinkable from existing—if needed, with force, legally and illegally. The stakes are high indeed, exceeding by far what a book can express.[17] It is my own geopolitical and academic entanglement that urges me to be a traitor *to* (while still expressing-with) language, the regime most prevalent in the fields I am trained in. While I am not making claims about what language *is*, I am interested in how to make-sense-*with* language, to sense and be sensible to words and concepts, without falling into either representationalism or empiricism.

Deleuze and Guattari's criticism of expression's reduction to the signifier and the imperialism of language rooted therein serves as an important entry point. For by thinking-with the dynamics they sense in Kafka's texts—namely, that each word metamorphoses by being opened up to "an asignifying intensive utilization of language"—they suggest that language can, and in fact has to, *mean* differently.[18] Precisely because language is arid, its symbolic, significant, or signifying usage has to be opposed by a "purely intensive usage" so that we "arrive at a perfect and unformed expression," which is "a materially intense expression."[19] This, however, has nothing to do with the biological determinism that has governed the relation between reason and language, nor does it advocate a relativism where everything could mean everything else. It instead requires a different understanding of matter altogether: Language ought to be encountered *matterphorically*.

This explains my second entry point: Barad's claim that matter and meaning are entangled such that a word cannot be a word and a concept cannot be a concept. What I wish to ask then is what the relationalities, entanglements, and dynamics *are* that relate, entangle, and dynamize—and what and how do they make-sense by keeping sense in a state of being constantly made and remade? While this question, by the very nature of what it touches upon, cannot be answered universally, I found that thinking-with the brittle star, offered as *matterphorical expression* by Barad and mattering-*forth* with every engagement, can guide thought toward making language itself a material expression, dissolving alleged representational units (words, concepts, images) and material entities (atoms, things). This in turn is a prerequisite for acknowledging that we do not yet know what thought can do, let alone what thought is doing. What is more, this is also the starting point for every *matterphorical case study*: investigating entanglements in their specificity and indeterminacy, where an immanently entangled world expresses itself *in* difference, where neither recognition nor identity can gain hold, where an electron does crash into a language. Reaching that point where electron and language are in touch and making that argument requires some detours: from engaging in collaborative thought to questioning what it means for concepts and words to be real. It involves challenging ontological assumptions by thinking-with matterphorical expressions that are sensitive to dynamics ungraspable by representational thought and that relate relations differently.

Let us return then to the brittle star. As pointed out, Barad's essay thinks-with and explicitly builds on Haraway's crucial feminist critique of disembodied epistemologies in "Situated Knowledges." Haraway's argument for the importance of situated, embodied knowledges is a critique of the powerful metaphor of disembodied vision in knowledge production and entails a call to switch to different metaphors.[20] Published in 1988, the text challenges universalized ideas of (scientific) knowledge that rest on an objective form of vision (God's trick—seeing everything from nowhere) and rhetorical tradition. Her call for a shift to metaphors that convey partial objectivity and embodied perspectivism is articulated in response to two conflicting positions on objectivity: social constructivism and empiricism. For Haraway, both positions—one claiming that everything is only text and thus constructed and the other stating that everything is empirical—are problematic. While the concept of situated knowledges and perspectivism takes on aspects of both, the call for metaphors implies a tendency toward the former. This must be understood in the theoretical and disciplinary milieu in which Haraway was writing in the late 1980s, which was especially, and importantly, concerned with the undoing of naturalized dichotomies. What is more, metaphors and language are powerful. As linguistics, literary theory, and cognitive science have shown, their influence on our modes of thought and action must not be underestimated. However, the sun is different from metaphors, as it is from the sun metaphor. What is more, no two sun metaphors are the same (for reasons not restricted to their semantic fields) and neither is what is directly referred to as the sensible sun ever the same. Haraway is clear about the insufficiency of metaphors as a means of thinking. The emphasis on material-semiotic entities and their embeddedness in more-than-human worlds that guides her later work allows us to understand not only her method but also her figures as matterphorical rather than metaphorical. In her *Companion Species Manifesto* published in 2003, for example, she clarifies right at the beginning what it means to think-with and theorize-with a dog: "Dogs, in their historical complexity, matter here. Dogs are not an alibi for other themes; dogs are fleshly material-semiotic presences in the body of technoscience. Dogs are not surrogates for theory; they are not here just to think with. They are here to live with."[21]

This also implies that even though Haraway uses the tool of language (the word *dog*) with all its histories (of thought, paper, cynology, text-processing software, trading routes and exchange rates, misunderstandings, legal terminologies, etc.) to reach out to readers, a dog is not to be understood as metaphor. It must not be reduced to standing for something else but rather has its own standing in the material and semiotic structures that produce and manifest meaning in the world. The claimed superiority of language (or particular languages) as the primary mode of expression for rational thought has a long history in colonialism. As Glissant explains, "the first thing exported by the conqueror was language," legitimizing attempts at domination and "culminating in the thought of the empire."[22] We have already seen (although in a different yet related manner) the danger inherent in the imperialism of language. We will see later how the need for a specific sensitivity to matterphorical expressions (and their relationality), a synaesethics, evolves against the hegemony of what Deleuze and Guattari call "State thought" and invokes what Glissant describes as "poetic thought"—alert and "beneath the fantasy of domination," seeking "the really livable world."[23] What is crucial here is that Haraway's concept of situated knowledges reintroduces an embodied, perspectival, and necessarily fragmentary account of objectivity in order to challenge the transcendent, disembodied, and universalizing ideas of objectivity, thinking, and knowledge production. However, where Haraway emphasizes the power of metaphor in producing and controlling meaning, Barad's essay questions the ontological assumptions underlying any dominant mode of meaning production.[24] And like Haraway, Barad also has to face the issue of making use of language while arguing against its hegemony, as well as that of representationalism in general. Another challenge we face when thinking-with the brittle star as matterphorical expression, rather than metaphorizing and displacing it onto a representational plane, pertains to the ontology of knowledge. As argued by Barad in their essay, attending to the brittle star's entangled mode of thinking and being challenges our conception of thought and knowledge production. However, as Barad also makes clear, the brittle star is not simply a metaphor. The imperative is not to think *like* the brittle star; a mode of thinking entangled with a mode of existence rejects orders of analogy and structures of resemblance.[25] *Like*—as linked to metaphor, allegory, and comparison—proposes a structural analogy of relations apt to describe stable entities, separable and bounded, at the expense of understanding processes of entangled becoming.[26]

If the brittle star is not a metaphor, and if what we encounter in the essay is what we understand as language (including the term *brittle star*), then what is it that we are thinking-with by attending to that essay? Or, how do we encounter language nonrepresentationally and does it, in fact, open up different modes of existence and livable futures?[27]

Comparison and analogy fail in many ways, across assumed scales and regimes of signification. First, if a word is not a word and a concept is not a concept, then what we encounter in the text as "the" brittle star is already multiple. What is more, the brittle star (concept or word) we encounter in this text—whether printed on paper, displayed on a digital screen, or read to us by a human or a computer-generated voice—is not a brittle star we would also, under certain conditions, encounter in shallow-water marine environments. In addition, we, in attending to Barad's text, also do not encounter the *same* brittle star (or the *same* text even). How would we even think a brittle star in its specificity, given that we approach it by means of a text and use the definite article ("the brittle star"), which, in its linguistic function, indicates a generalization? Does the use of language and its referential function not inevitably transform the brittle star into an idea inherently different in kind from the brittle star in shallow sea waters—into a rhetorical device of comparison and suspension, devoid of its particularity? What these questions show already is not only the difficulty of expressing nonrepresentational relationality while using language but also that when refusing to think multiplies meanings analogically, *things* seem to make less sense. Or, put differently, sensemaking no longer pertains to things, ready-made and with clear boundaries. The brittle star becomes many, each of them multiple, to be approached as entangled relations rather than bounded entities of (shifting or fixed) meanings, ready to be related to other representational and material entities. To a certain extent, although still committed to representational modes of thought, this has also been argued by poststructuralist theories focused on language. However, matterphorics departs here. It works under the assumption that everything is real, rejecting the mind and matter dualism and attending to the specific entanglements of meaning and matter that a *matterphorical expression* (be it a word, a crystal, or a movement) makes graspable.

Because of this departure, the first question to address in order to demonstrate how the brittle star can be understood as matterphorical expression pertains to the ways in which the question of realness, especially when it comes to language, has been posed. Here, Deleuze and Guattari's work (and also and especially the latter's solo work) on breaking the dualist structure that separates representation from the real is helpful. They are, of course, not alone. Postcolonial and feminist-scientist scholars, for example, have repeatedly stated that the search for and subsequent exclusionary claim of "the real"—the real brittle star, the original—are a territorializing, colonial practice that establishes otherness and renders it a copy, inferior and inauthentic. Haraway specifically has argued that practices of knowledge production that "only displace the same elsewhere, setting up the worries about copy and original and the search for the authentic and really real," are practices based on reflexivity and reflection. Even our "critical practices," she laments, are mainly based on reflexivity.[28] Needless to say, if what *is* real can only be understood as *the* real, leading to both the possibility of its negation and the naturalization of its affirmation, then it is understandable that theory, especially critical and poststructuralist theories, fears any proximity to what is considered real—matter, world, existence, life, nature, physical forces, chemical and biological processes. While I agree that these dangers, including that of biological determinism, must be taken seriously, I do wish to question the assumption of the inextricability of what is called "the real" and the inevitable naturalization of meanings. Turning away from matter—and thereby from ontological questions such as what matter *is* and *how* matter comes to matter—does not deprive matter of meaning. Making sense remains matterphorical and inseparable from political, social, and legal phenomena. The task at hand is neither to attempt to declare modes of representation superior to or detached from matter(s) nor to declare representation (as such) real. It is rather to show how that which is mistaken for representation is in fact an entangled *matter* and therefore real. To reiterate, this is why, for Deleuze and

Guattari, "the abolition of all metaphor" is a prerequisite to understand that on the plane of immanence, "all that consists is Real."[29] Real, however, denotes neither fixed objects and terms nor representability. The plane of immanence is not a plane of reference.[30] "We fall into a false alternative if we say that you either imitate or you are," they argue, for what is real is becoming itself.[31] The dynamism at play here is not the Cartesian cut but what Deleuze and Guattari express as "double articulation" (producing not only A-B but also AB, where A, B, and AB are themselves doubly articulated, and so on). It is worth noting here that in its dynamic and by aiming to challenge the ontology and epistemology of matter-meaning-making, this resonates strongly with Barad's intra-action as a cutting-together-apart. The cutting-together-apart, in terms of double articulation, lies in the co-constitution of "binary relations" and "biunivocal relationships obeying far more complex laws."[32] In *A Thousand Plateaus*, this specific dynamism is thought with not the brittle star but the lobster, "a double pincer, double bind," a "multiplicity of double articulations affecting both expression and content."[33] Aiming to be attentive to dynamisms aside from the Cartesian cut, doing theory matterphorically does not seek to explain or address "the Real." As everything, without being a thing, *is* real, matterphorics takes seriously the complexity and response-ability that arises from that shift in thought.

Barad's essay—and general philosophy—runs along these lines too. For if we follow their agential-realist argument that "knowing, thinking, measuring, theorizing, and observing are material practices of intra-acting within and as part of the world," then no analogy and no metaphorical *like* that links ready-made entities can gain hold here.[34] As stated above, while Haraway importantly criticizes the disembodied and detached observer, argues for an embodied and partial perspective, and calls for different metaphors, Barad pushes these arguments in a different direction by challenging the ontological assumptions underlying representationalism in physics as well as sociopolitical thought. They achieve this by focusing on the "material-discursive nature of boundary drawing practices" that produce meaning in the form of differences that *matter*, which, when understood as onto-epistemological claims, make it difficult for representational thought to take hold.[35] First of all, according to Barad, no inherent subject-object distinction exists. The primary ontological units are phenomena, which do not merely mark the epistemological inseparability of "observer" and "observed" (which is, roughly speaking, the argument made in "Situated Knowledges") but materialize the "ontological inseparability" of intra-acting (and therefore entangled) agencies.[36] Phenomena, including space and time, are, in other words, not *in* space and time but "ontological entanglements" that "extend" across spaces and times. Needless to say, Barad's claims, whether accepted or not, have significant implications for both quantum physics and philosophy. In terms of what they mean for modes of sensemaking, they importantly advocate for a reconceptualization of both objectivity and referentiality. Objectivity, rather than relying on an absolute condition of exteriority, is to be understood as "a matter of exteriority-within-the-phenomenon."[37] No cut, no act of boundary drawing (discursive practice), is enacted once and for all. No-thing, Barad argues, "stands separately constituted and positioned inside a spacetime frame of reference, and no divine position for our viewing pleasure exists in a location outside the world."[38] Accordingly, there is no observer-independent referent (object) in the word, awaiting signification. Rather, the referent is a phenomenon. Different agential cuts—resolving ambiguities not once and for all but only within a phenomenon, for a given context—produce different phenomena.[39] Objectivity, in taking into account and being accountable for the specific exclusions enacted by cuts, is "about being accountable and responsible to what is real."[40] The task is therefore not to look differently *for* or *at* different objects but to shift the notion of objectivity itself in order to response-ably think-with and make-sense-with an immanently entangled, material world.

This leads to the second point—namely, how matter comes to matter. While Deleuze and Guattari state in *A Thousand Plateaus* that "unformed matter is not dead, brute, homogenous matter, but a matter-movement bearing singularities or haecceities, qualities, and even operations," thereby demanding that dominant epistemological assumptions be confronted with their underlying ontological ones, it is Barad's work that investigates and challenges the latter in great detail.[41] According to Barad, *matter* is "substance in its intra-active becoming," and as such it is not situated *in* the world but is rather "worlding in its materiality."[42] It is not a thing but a doing.[43] Precisely because, according to agential realism, the "primary ontological units are not 'things' but phenomena" (entanglements, (re)articulations of the world) and the "primary semantic units are not 'words' but material-discursive practices through which (ontic and semantic) boundaries are constituted," meaning and matter are not separate.[44] Rather, they are entangled, cut-together-apart; that is, co-constituted by agential cuts that do not operate according to subject/object logic. Meaning is not a property of words (or other assumed representational entities) but "a performance of the world in its differential dance of intelligibility and unintelligibility."[45] This also implies that words and concepts are not representational units but entangled expressions of matter and meaning production. They are real yet not the *same*; cuts *matter*.[46] Representational thought, however, constitutes concept *and* word as representational units, constructed by cognition and signification, and considered separate from the material world. It also facilitates the mistaking of one for the other and thereby leads to concepts being untethered from their specific entanglement qua words (themselves also subject to such untethering) and applied to a different context (or rather, onto-epistemological field) by means of analogy. The *same* concept becomes the *same* word, inhabiting a syntactically correct spot in another sentence, another book, about another object, on another subject.[47] Elizabeth Povinelli emphasizes that the violence (not simply erroneous nature) of such mistaking is usually not felt by those juggling with representational units (in specific academic spaces, for example), which facilitates the perpetuation of this habit of thought and knowledge production. Thinking-with Povinelli and our cowritten piece, however, we see that concepts, including their linguistic expressions, *matter* and "matter-forth in material bodies and relations," redirecting "matters of meaning by changing the dynamics of workability in a region within matterphorical fields."[48] What is more, we see that words, even if presumably the *same* ("brittle star" and "brittle star"), are neither the same nor ever just words: "As if a word could be a word!"[49] And thinking-with Barad, pushing what a concept can *be* even further than Deleuze and Guattari did, we can understand concepts as "specific material doings or enactments of the world," as *spacetimemattering*, expressions of the world "doing theory (theorizing)."[50] Thus, Barad affirms in our conversation that concepts have "to be understood matterphorically" alongside all their entangled expressions in their specific entanglements[51]—words, sounds, colors, quarks, encounters, events, breaths, turnings, attractions, powers, forces, a lobster meeting a brittle star for the first time without having agreed on a sign (let alone a social contract): *matterphorics*.

Taking seriously the meaning of these onto-epistemological shifts and their implications for nonrepresentative yet meaningful encounters with language, we can reemphasize that there is indeed no point in asking which brittle star is "real," or how the words "brittle star" relate to the concept "brittle star" or to a brittle star at the sea bottom. None are fixed (whether in terms of matter or meaning). All are real, all are cut-together-apart, and nothing can *be* the *same*. Thinking-with is a *synaesethical* practice. It requires links and relations that have not been preconceived. Thinking-with a brittle star in the ocean, let alone a specific one, does not make-sense matterphorically without us getting our feet wet and holding our breath; probably also not without learning how to swim, which, like learning a foreign language, means, as Deleuze writes, "composing the singular points of one's own body or one's own language with those of another shape or element, which tears us apart but also propels us into a hitherto unknown and unheard-of world of problems." What are we dedicated to if not, he asks, "those problems which demand the very transformation of our body and our language?"[52] And why should not learning to swim a language, touch a word, or drink a concept (the concept of toxicity, for example) become unknown?

Besides, what does it mean to encounter *in* and *as* representation? What kind of intimacy does not even assume the possibility of being, or having ever been, in touch? And what kind of distance rests on an absolute ontological cut? While various modes of doing theory, including critical theories, have developed an awareness of the problematic nature of making comparisons between original and copy, the often alleged representational proximity (for example, the proximity between an analogue and a digital version of a particular photograph or the use of a word in two different contexts), which rests on the separation of matter from its assumed image, persists. Why is it, for example, assumed that the words "brittle star" as they appear in a text that is printed out by two different devices, in two different geopolitical locations, at different times (perhaps one at night and the other during the day), on different sheets of paper, from different kinds of wood, (and, and, and . . .), can have the *same* meaning, while the distance between a specific *Ophiomastix wendtii* navigating through the shallow waters of the northeastern Brazilian continental shelf and the word used in the text is assumed to be much more significant, almost unbridgeable? Expressions are multiple, their onto-epistemological relationality an open question(ing). A brittle star sliding over the muddy sea ground: an infinite number of singular grains of sand, quartz crystals, molecular structures composed of elements, silicon and oxygen atoms, protons, neutrons, electrons, virtual particles, forces that bind and separate, move and put to a halt. Sea water, varying degrees of salinity, chloride, sodium, natrium, magnesium, calcium and potassium, inorganic and dissolved organic substances, atmospheric gases, protons, neutrons, electrons, virtual particles, waves of all kinds mingling, vortices, forces that move and resist movement. A brittle star rejects its arm. Where does a brittle star begin and end? What and how *is* distance and proximity? What makes an arm? A brittle star finds a textual or photographic expression in millions of pixels on a digital screen through the altered alignment of molecules and the optical properties of liquid crystals encountering light and an electric field, protons, neutrons, electrons, virtual particles, forces of attraction and repulsion, movement and rest. Another arm reaches across the paper page: histories of wood, trees without arborescent structures, cutting forces that do not separate. Each pixel, each particle, each wave, each force carries with it all its (material, economic, social, political) cuts, relations, and encounters. Questioning

Enlightenment by diffracting sunlight: different wavelengths, resistances, and encounters, moving differently through history and atmospheres, reaching out and touching the light-sensitive cells scattered across the skin of a brittle star. How can representational thought, how can comparison and analogy, possibly and response-ably make sense? What is sacrificed, rendered negligible, for the ever-same sense that is and has already been in the making for centuries now? How many histories, how many molecular and social bonds, how many complicities of force and power have been declared insignificant, excluded from mattering? How can we still claim to know that what is deemed negligible in the course of analysis has less weight than the brittle star—and should not the impossibility of answering that question, regardless of its usefulness, make us rethink our assumptions about both matter and mattering?[53]

## Difference and Mattering

Attending to these questions is an ethical and political concern. It is a matter of brittle stars as well as of the legal and political concepts we live by—among others, democracy, legal personhood, force of law, contract and consensus, equality, justice. We only need to imagine another concept, practice, or word that is taken for granted, definable or defined, measured, delineated, and eventually represented in and by law. What if we had thought-with breathing or breath? Do we *all* know what it means to not be able to breathe, do we all *know* what #icantbreathe means? How does the law know? Does (and can) difference matter?[54] Although it will be the second part that demonstrates what a matterphorical case study can do with the law, I wish to at least point to the importance of difference as nonoppositional and inextricable from the aforementioned articulations of the world. This understanding is a prerequisite for breaking with habits of representational thought. The question that makes it impossible for me to skip over difference is the following: If we refuse to designate analogy and comparison as privileged modes of sensemaking, if we accept that every *like* has to be "made into an expression of becomings instead of signified states or signifying relations"[55]—that is, if we reject the notion of language as signifying and purely representational—then what *is* it that relates that which relates, expresses that which expresses, thinks that which thinks, and differs that which differs?

The question of what difference *is*—what ontology underlies it, whether other notions of ontology are possible, and if so, what potential they hold to change legal theory, systems, and orders—is crucial to me. This is certainly also owed to the thinkers and thoughts I think-with. As early as 1954, in a review of

Jean Hyppolite's *Existence and Logic*, Deleuze argues for an "ontology of difference" that articulates a "theory of expression where difference is expression itself," and sense is made (or makes itself) differently.[56] Barad's concern, articulated many years before *Meeting the Universe Halfway*, is the "nature of the interplay of the material and the cultural in the crafting of an ontology," which in their later work becomes "a relational ontology that rejects the metaphysics of relata, of 'words' and 'things.'"[57] However, Barad does not offer an ontology as a totalizing claim of what the world *is*. Rather, according to their agential realist account, "ontology is the 'theorizing of what is' by materializing things in certain ways, a particular form of intra-acting, and as such part of the world."[58] The world, put differently, "*is* intra-activity in its differential mattering."[59]

I argue that, reading Deleuze's differential ontology, which attempts to challenge the strand of Western philosophy that is complicit in grounding regimes of power in identity and representation, together with Barad's onto-epistemology, which aims to challenge the violence of representationalism that is complicit in sending missives and missiles, allows for an *onto-epistemology of difference*. Difference, in that view, comes to matter as an entangled expression of the world rather than as a form of negation or the result of comparison in a secondary manner. An *onto-epistemology of difference* does not allow for Cartesian dualism and urges us to find modes of sensing and sensemaking through the ongoing questioning of *matter*ing in its entanglement. To state the obvious, Barad's and Deleuze's (as well as Deleuze and Guattari's) notions of difference are not only not the *same*, their concepts are not comparable (which is also an argument this section makes), and attempting to approach their work by means of comparison would run counter to their philosophies. Besides, and relatedly, my method is expressly not comparativist. In what follows, I attend to a particular dynamism, sensible in the work of the thinkers mentioned, that neither allows for a Cartesian cut nor makes a final claim about what the world *is*. I find that this dynamism, as one not of language but of the world, is most precisely captured by the *plane of immanence* (Deleuze and Guattari) and the *void* (Barad). I wish to sense this dynamism, expressive of an onto-epistemology of difference, with the brittle star, making thought pass through all representational barriers.

Granted we accept what I have suggested so far, let me first restate the problem: No fixed entity exists as such, becomings are real, everywhere there are cuts that cut meaning and matter together-apart. A brittle star: quarks, calcium and potassium, liquid crystals, modes of reading, C++, software license, ebbs and flows of stock market price, tenure track position, tidal waves, movement, turnings, forces. In this, Guattari's outcry comes to mind: "Same mess all over again. I'm so jealous of your ability to organize and classify things!"[60] Of course, Guattari plays with irony, expressing the unavoidable challenge of approaching sensemaking as a process and practice independent from representation. However, the sentiment is shared. If no *thing* makes sense, does *nothing*? (In fact, as Barad shows, it does!) Where to begin? And where to stop? How far, in terms of an ethics of thought, does thought have to go? How far can a concept be pushed before its contours become brittle? How much pressure can it stand before it is torn apart? How many molecular bonds does it take to prevent a word from transcending the world, how much power is needed to overthrow representational regimes of meaning? Deleuze and Guattari, in articulating the "plane of immanence" (often also "plane of life" or "plane of consistency"), offer a (non)concept, as it is not thought but the "nonthought within thought," or that which "must be thought and yet cannot be thought,"[61] that aims to address precisely these questions. What is crucial here is that, on the plane of immanence, "the most disparate of things and signs move upon it: a semiotic fragment rubs shoulders with a chemical interaction, an electron crashes into a language, a black hole captures a genetic message, a crystallization produces a passion, the wasp and the orchid cross a letter."[62]

If an electron crashes into (a) language and a wasp and orchid cross a letter, then combustion can set a concept on fire and thought can be attracted (or repulsed) by electromagnetic force. But what *is* the plane of immanence that makes representation impossible? Many answers have been proposed.[63] At times, Deleuze and Guattari's descriptions seem inconsistent, shifting, and sometimes contradictory. This is why my interest lies not in providing a descriptive account of a concept affixed to specific terms ("plane of immanence") but in being attentive to how matter and meaning are related by a dynamism that they aim to express by means of a *matterphorical expression* (plane of immanence). In short, the virtual yet real plane of immanence is described by Deleuze and Guattari as the "unique plane of life," characterized less by a unity of substance than by an "infinity of the modifications that are part of one another."[64] It is to be understood, they argue, as a plane of univocality, opposed to analogy precisely because *being* (or Difference, the One, pure immanence) "expresses in a single meaning all that differs."[65] There is neither identity nor sameness on the plane of immanence. Rather, the "most disparate of things and signs" move upon it: There is "no 'like' here, we are not saying 'like an electron,' 'like an interaction,' etc."; these are "electrons in person, veritable black holes, actual organites, authentic sign sequences."[66] Importantly, however, Deleuze and Guattari emphasize that there are nevertheless "rules." The plane of immanence "is in no way an undifferentiated aggregate of unformed matters, but neither is it a chaos of formed matters of every kind." What underlies it is a dynamism ("abstract machine") bringing about a "continuum of intensities, combined emissions of particles and sign-particles, and conjunction of deterritorialized flows."[67] It is here that I wish to turn to and think-with with Barad and their quantum-field-theoretical interpretation of the vacuum—that is, the void or nothingness—of which they write that "there are an infinite number of im/possibilities, but not everything is possible."[68] Or, described differently, "the vacuum isn't empty, but neither is there any/thing in it."[69] In "What is the Measure of Nothingness," Barad aims to formulate the dynamism of the void, offering their interpretation of virtuality and vacuum fluctuations. They convey the void (which is, according to quantum field theory, not separate from, and according to Barad intrarelated with, particles and field), as "the quiet cacophony of different frequencies, pitches, tempos, melodies, noises, pentatonic scales, cries, blasts, sirens, sighs, syncopations, quarter tones, allegros, ragas, bebops, hiphops, whimpers, whines, screams . . . threaded through the silence, ready to erupt, but simultaneously crosscut by a disruption, dissipating, dispersing the would-be sound into non/being, an indeterminate symphony of voices. The blank page teeming with the desires of would-be traces of every symbol, equation, word, book, library, punctuation mark, vowel, diagram, scribble, inscription, graphic, letter, inkblot, as they yearn toward expression."[70]

The question of expression, thought-with and by the void, is an ongoing questioning of "the very im/possibility for non/existence."[71] It has nothing to do with representation. Indeed, for Barad, expression does not fall into a mind/matter dualism. An electron *does* crash into (a) language. And lightning, long before it strikes, undoes Western metaphysics, making a difference, yearning for expression: *Enlightningments*.[72]

Refusing to accept that sensemaking demands signifier and signified, always appropriately out of touch, requires pushing thought to think what it cannot think. For Deleuze (and Guattari), this means encountering what is sensible yet not recognizable, while Barad describes it as going "all the way down" (meant in an ontological rather than spatial or temporal sense) to how being (and nonbeing) is expressed intra-actively. What thought encounters if being traced (and by tracing) "all the way down" is no/thingness—a desire of and for being expressed, "a lively tension, a desiring orientation toward being/becoming."[73] Since all is *of* nothing (as no thing, not even space and time, preexists), expression (understood via the underlying dynamism of intra-action or double articulation) is never secondary to existence. Now, thinking with Deleuze and Barad, the question is what this means for difference if difference itself is not a secondary phenomenon or concept either. In *Difference and Rep-*

*etition*, Deleuze states that "difference is behind everything, but behind difference there is nothing."[74] And, as we learn from Barad, nothingness is always already expressing (itself as) "a field of differencing (differencing entangling)" flush with the yearning to be/come, that is, toward existence/existing.[75] Difference then is neither negation nor the result of comparison but, expressed as the ongoing material articulation of the incomparable desire to exist, an entangled process of mattering (in its double meaning) and, as such, expressive of an immanently entangled world. This comes with radical ethical implications. For if, as Deleuze claims, "each difference passes through all the others," and if, as Barad writes, differentiating is not about othering/separating," as even otherness "is an entangled relation of difference," then our dominant modes of analysis and thought (including legal thought) are either unconsciously inadequate or deliberately ignorant of not only the dynamics of the world but also of the many modes of existence rendered unthinkable by means of allegedly being different *from* some*thing*: from what is considered significant, of value, original, legitimate.[76] Reason might not be an ontological concern in claiming the power to speak the first and final judgment about what and how the world is. However, unreasonableness, as that which must neither exist nor be thought, certainly is.

What and how does all of that (or, in fact, anything) mean, then? Where does thought begin, and is there an end? Can thought come up with a conclusion, once and for all? If thought cannot take its usual hold, is there still a point in thinking, let alone critical thinking? Addressing the challenge that a rejection of representationalism poses to our notions of thinking and sensemaking, Barad clarifies that there is no endpoint when it comes to tracing entanglements. Yet this does not mean that no sense can or should be made. In fact, the process requires close attention to how meanings come to matter. Precisely because there is no endpoint, what we need to understand, by paying close attention to how differences are constituted and enfolded, how they matter, and to whom, are "diffraction effects."[77] Lest difference is mistaken for simply the outcome of dualist modes of thinking, Barad cautions that *"diffraction is not merely about differences, and certainly not differences in any absolute sense, but about the entangled nature of differences that matter."*[78] Through the matterphorical case study on a man falling from the stratosphere, I will demonstrate how "going all the way down" and tracing relationalities is a creative approach aiming to make-sense-*with* differently. As is also the case with the brittle star, each investigation, each attempt to think, must take seriously the entanglement of meaning and matter, tracing relationalities and complicities from a specific entry point, a specific cut, outwards (without assuming absolute exteriority or declaring a center).

Thinking is always thinking-with. We could have thought with a lobster, a specific gesture, the intake of a breath (or its impossibility), an electron, a taste of bitterness. What I aimed to show, by thinking-with the brittle star, drawing upon the work of Barad and Haraway (among many others), was the possibility and need for a *synaesethics of thought* that resists modes of representational thought and yet does not fall into empiricism. Representationalism makes modes of existence thinkable and renders them thoughtless. Empiricism, so far, has not escaped its ontological assumptions. Whenever thought recognizes something, it ought to trace further, think what is yet unthinkable, learn to sense difference, make-sense without mediators, until it reaches a point of indetermination where it can make-sense-*with* differently. From there, it can think-with and theorize anew—with what has been unthinkable, unrecognizable, unrepresented, and unrepresentable. For Barad, theorizing is a "form of experimenting" that is by no means a privilege or exclusive capacity of human beings. Rather, "theories are living and breathing reconfigurings of the world." Ontological thought experiments do not follow the logic of the sign.[79] If theorizing is not an exclusively human endeavor, thinking is inextricable from being, and being is not a settled matter, then thought is not necessarily, let alone predominantly, expressed in language that in turn cannot be the indicator of its legitimacy. In fact, as we will continue to see in the next chapter, thinking *is* nonrepresentational. Matterphorics acknowledges that in times like these, theory must (finally) lose its mind so we can learn to become response-able and sensitive to what is not yet thinkable, to what at first might even seem unthinkable and nonsensical: an electron crashing into (a) language, a Red Bull logo breaking through the sonic barrier, oxygen molecules pushing legal theory, life and law becoming cutting edge.

# 4

# Unreasonable Canons

## LAWS OF THINKING

He is not the master of what is thinking in him. He must restrain himself, and will only be considered cured when he has mastered this art. But how should he be cured from that which makes him rather derange the law than subject himself to it . . . to the mad, invalid law. —CHRISTA WOLF, *Kein Ort. Nirgends*

Always obey. The more you obey, the more you will be master, for you will only be obeying pure reason, in other words yourself. —GILLES DELEUZE AND FÉLIX GUATTARI, *A Thousand Plateaus*

I had the opportunity to go to Greece, which was amazing. So that was the fall break of my sophomore year, after I had completed the humanities sequence; and going to the origin of Western culture—because the humanities sequence starts with Homer and the Illiad—that was really wonderful . . . because you get to see, this is the place that inspired these thinkers that are the foundations of why I think the way I do. The humanist sequence, and the humanist community, is something that is known on campus, that other professors respect; it is something that tells them something about who you are, when you tell them that you have done it. —UNDERGRADUATE STUDENT, Princeton University

*"This is the place that inspired these thinkers that are the foundations of why I think the way I do."* This statement was made by an undergraduate student in an advertisement video for the so-called "Humanities Sequence," which is described on the website of Princeton University's Humanistic Studies Program as "an intensive year-long introduction to the landmark achievements of the Western intellectual tradition" that is "enhanced by trips to museums, plays, concerts and art galleries both on campus and in New York City." The only prerequisites for taking the sequence, the website further states, are a "love for reading" and an excitement for encountering "beautiful books," as it "attracts a self-selecting group of students who are ambitious, dedicated, and willing to work hard."[1]

I have decided to open this chapter with a student's quote not to criticize, let alone expose, the student but because the statement (which was a reflection on the sequence after having completed the year-long class) speaks very clearly to the core issues raised in this part of the book: where and how thinking can take place (and where it takes *a* place); the idea of an origin of thought; attempts to remain truthful to, repeat, or retrieve the origin; the power of intellectual foundations and how they persist over time; the figure of the ingenious thinker; and, finally, the possibilities and limits of thinking-*with* and *in* a particular tradition of thought. In addition, the statement highlights both the importance of recognizability and the presuppositionality of what is already known. It reveals the acceptance of an authority (be it the canon or a respected professor at an Ivy League campus) and, in relation to that acceptance, a tacit understanding that such apprenticeship will lead from mastering (the canon) to becoming-master oneself: "It is something that is known on campus, that other professors respect; it is something that tells them something about who you are, when you tell them that you have done it." Finally, the fact that this articulation takes place in a specific institutional setting, one that is centered on thinking and knowledge production, foregrounds the importance of the questions raised—though they are neither exclusive to this institution nor limited to a specific discipline.

Alfred North Whitehead, whose work is currently not on the Humanities Sequence's reading list, writes in his 1944 publication *Modes of Thought* that "the task of a university is the creation of the future," at least "so far as rational thought . . . can affect the issue." He continues: "the future is big with every possibility of achievement and of tragedy."[2] Stated as such, Whitehead's claim might not seem too far from what humanistic core curricula aim to do, at least insofar as both rational thought and the classical humanities are grounded in the dogmatic and moralistic image of thought. However, by looking closer at his altered concept of reason and uncovering that "rational thought" might not be what we assume it to be, we see that it is a different mode of thinking entirely—not just an altered philosophical canon—that carries the potential to curtail the tragedies produced by rational thought.

Indeed, in a set of lectures Whitehead delivered at Princeton University in 1929 entitled *The Function of Reason*, he articulates a different notion of reason, one that is not a "godlike faculty which surveys, judges and understands," but rather intrinsically organic in its origin and purpose. As I wish to demonstrate, rational thought guided by this notion of (speculative) reason does not aim to transcend the world or to colonize terrestrial and epistemological spaces. In fact, reason is so intimately entangled with life and with the physical world, with its movements and processes, that thought too can be something utterly different: sensitive in unexpected yet not random ways to its being of the world.

This concept was radical at the time of the lectures' delivery and it still is radical, given the dominance of the image of thought that relies on recognition, identity, and representation. It is also against this background that Whitehead states already at the beginning of his first lecture that the function of reason is "to promote the art of life," which involves a sensitive engagement with the entire lived world rather than the logical structuring of propositional knowledge.[3] In opposing a concept of reason that asserts itself as "above the world" and seeks a "complete understanding" to reason "as one of many factors within the world" that seeks for "an immediate action," Whitehead formulates a theory of thought that is both immanent (of and within the world) and, as Erin Manning has it, in the in-act.[4] Most importantly, reason, as Whitehead understands it, is not a cognitive faculty that executes a preexisting order to govern thought. It is the opposite: Thought contains anarchic elements and is oriented toward novelty and difference in and as life. The anarchic elements and the importance of difference as novelty rather than a relation of opposition or similarity will be also crucial for Deleuze to break with a mode of thinking that, for him, had remained unquestioned. Moreover, Whitehead's reason—his notion of speculative reason—can be seen as a possible response to Deleuze's inquiries about how the earth thinks. Not only does it affirm that the earth does indeed think and elucidate how we conceive of its practices, it also challenges the progress narrative inherent in the Enlightenment version of reason. For, as Whitehead says at the end of his preface, reason is "the self-discipline of the originative element in history. Apart from the operations of reason, this element is anarchic."[5] It is worth more closely examining what exactly Whitehead means by reason, because it enables a mode of nonrepresentative thought, constructive of matterphorical expressions, attentive to an ethics of sensing and sensemaking, and as such, it is simply not applicable *to* what is considered a given (world, context, time, space . . .).

## Reasonable Matterphorics

For Whitehead, the function of reason is "to promote the art of life" consisting of the threefold urge to live, to live well, and to live better. Whitehead does not consider this urge to be individualistic because living is too intimately related to all processes in and of the world and to all of its virtual and actualized modes, expressions, and embodiments. It is a fallacy, he says, to believe that fitness for survival, as Charles Darwin argued, is equivalent to the best exemplification of the art of life because "life itself is comparatively deficient in survival value."[6] The art of persistence is, after all, to be dead. The art of life, then, has to do with the introduction of novelty—i.e., difference—that prevents life, the organism, and even thought from existing in mere repetition. Static life never truly attains stability, only fatigue and consequent slow decay.[7] Novelty, on the other hand, strives for living—even living well and better, where what is meant by "well" and "better" depends on the single occasion, the actualized entity, rather than on moral or universal values. For Whitehead too (although from a different perspective from how Deleuze, Foucault, and Barad approach this problem), the relation between empty space and thought ought to be reconsidered. He argues that it is crucial to abolish "the notion of valueless, vacuous existence." Vacuity belongs to abstraction and is "wrongly introduced into the notion of a finally real thing, an actuality," as it is assumed to be passive matter on which time inflicts change. Actualities, however, are never vacuous, only propositions and universals are.[8] As Catherine Keller puts it, "any ontological individual—particle, molecule, cell, each animal composed of them—is read as an actual occasion constituted of its relations to all its others" that "enfolds its universe and unfolds it differently."[9] What is more, each actuality is also always dipolar: It is "mental experience integrated with physical experience."[10] This is also why, for Whitehead, "whatever exists, is capable of knowledge in respect to the finitude of its connections with the rest of things."[11] Despite the common use of "mentality" in relation to *consciousness* (or *mind*), it is important to note that consciousness is not a necessary element of what Whitehead understands as "mentality." Mentality brings novelty into the appetitions of mental experience; it becomes self-regulative by canalizing its own operations according to its own judgments. This is where, for Whitehead, reason appears; namely, as "higher appetition which discriminates among its own anarchic production."[12] As such, reason too contains an anarchic element; it does not work according to pregiven structures but operates instead within the specific space-time-matter composition with which it arises. This is a crucial claim because it stands in strong contrast to the Enlightenment concepts of both reason and rational thought. Reason, as Whitehead defines it, is not the operation of theoretical realization, that is, of applying a theoretical and juridical system to the world. It is also neither universal nor a mode of apprehending pregiven orders. Rather, reason evolves from (within) and engages with a field of onto-epistemological difference(s). Life, for Whitehead, means novelty and difference. Indeed, reason functions here as a mode of sensemaking from *within*: from and with the world, from and with a specific set of relations, matters, embodiments. Without the anarchic appetition there would only be mere repetition. However, it is repetition (into which even anarchic appetition can descend) without novelty (i.e., without difference) that Whitehead's understanding of reason as a self-regulating, embodied, and embedded dynamism challenges. This is what Whitehead means when he concludes that this concept of reason is "the special embodiment" of the "disciplined counter-agency which saves the world."[13] As Manning describes, for Whitehead, nature *thinks* and creates thought as a "thinking in the event."[14] The force of thought, she writes, is immanent to the event and "is never thought as that which lands onto the event from outside its concrescence." Rather it is "the reason of nature, in nature, a concern with the very edges of the thinkable in its nonalignment to consciousness."[15] This also means that an encounter with what is unknown does not evoke a process of reasoning that evaluates, judges, and categorizes from a transcendent realm or position detached from the world, applying its law onto it. And, reminiscent of Barad's ontology of knowledge, Manning's characterization suggests that knowing implies an entanglement and becoming with the world.[16] Whitehead's speculative reason is less a counterconcept than an attempt to express a different onto-epistemological dynamism, addressed matterphorically and from a process-philosophy standpoint. It is neither moralistic nor prone (although certainly also not immune) to epistemological colonialism. Its mode of capture—the ordering function of reason—is inextricable both from its own anarchic, material, and entangled field of meaning and from other expressions and entanglements (a term that Whitehead does not use however) of the world. Reason, therefore, can neither create nor incorporate what is perceived as unknown into a structure of universal laws and the hierarchies these laws establish. Rather, reason "judges beings from within,"[17] something that is crucial for Deleuze and Guattari's ethics of immanence as well.

By contrast, not only does thought as reinforcement of the dogmatic image of thought (that which grounds epistemological laws) take (a) place but these taken places in turn also occupy thought, determining what it means to think and to produce knowledge. "When knowledge becomes a legislator," Deleuze cautions, "the most important thing to be subjected is thought"; namely, "to reason and to all that is expressed in reason." This has consequences for life as an open concept expressing the potentiality and power of existence: "Rational knowledge sets the same limits to life as reasonable life sets to thought; life is subject to knowledge and at the same time thought is subject to life."[18] In law schools, legal thought—a type of thought that relates most apparently to questions of the legitimacy of life, nonlife, and death—is tightly defined and replicated. At universities in general, cultural, political, and scientific thought is shaped and inscribed into neat narratives of the past and the future. As Rauna Kuokkanen reminds us, universities, with their histories in European Enlightenment values, still "reflect and reproduce epistemic and intellectual traditions and practices of the West through discursive forms of colonialism."[19] If the task of a university is, as Whitehead claims, to create a future by means of thought that is capable of affecting it, then the question of whether or not this will be a tragedy is inextricably tied to what the logic and space, the matter and temporality, of this kind of thought *are* and what it can *actually* do: precisely because the practices of thinking-with and thinking itself are *immensely material*. Whitehead is careful in using the indefinite article (*a* university) rather than the definite, which indicates that there might be as many futures, perhaps as many worlds, as there are universities and as many universities as there are modes of thought and practices of knowledge production that affect our futures and worlds. This is, Kuokkanen reminds us in her book *Reshaping the University: Responsibility, Indigenous Epistemes, and the Logic of Gift*, why "the academy will have to acknowledge that it is founded on very limited conceptions of knowledge and the world" and that it "has been guilty of (among other things) ignorance, erasure, silencing, and appropriation and theft of knowledge and culture."[20] Universities as we generally refer to them were established on the "principle of reason" as defined by late nineteenth-century German philosophy; they were both initiators and products of the intellectual and societal developments in Europe that included imperialism and colonialism. The university, as an institution, still derives its legacy and intellectual traditions from that particular historical and geographic context, and as long as the academy fails to renounce its "epistemic ignorance," it will be unable to acknowledge the "multiple truths" and, consequently, the multiple possible futures for shared knowledges and modes of living.[21]

In the same book where Whitehead states that the university's task is, as far as rational thought (which for him is, as we saw, onto-epistemological, in-act, and in relation to and even of the world) is concerned, the creation of a future, he also notes that "a philosophic outlook is the very foundation of thought and of life" because "as we think we live." Thus, decisions about what sorts of ideas we attend to and what sorts we neglect *matter*; this is why, for Whitehead, "the assemblage of philosophic ideas is more than a specialist study."[22]

In 1997, sixty-eight years after Whitehead proposed a different concept of reason and fifty-three years after he called attention to the significance of the kind of philosophical ideas we attend to, the South African writer J.M. Coetzee delivered the prestigious Tanner Lectures on Human Values at Princeton University. He decided, by reading to the audience a novella he wrote for that occasion, to give the stage to a fictional character, an eminent writer by the name of Elizabeth Costello. Reason, Costello declares in a talk she is delivering at an American college, "looks to me suspiciously like the being of human thought, worse than that, like the being of one tendency in human thought. Reason is the being of a certain spectrum of human thinking." She concludes: "And if this is so, and if that is what I believe, then why should I bow this afternoon and content myself with embroidering on the discourse of old philosophers?"[23]

## Nonrepresentational Image of Thought

Keeping in mind Whitehead's concern for rational thought and the university as a site of knowledge production, Kuokkanen's call for openness to various kinds of knowledge, Coetzee's giving the stage to Elizabeth Costello's question, and the undergraduate student's statement, I wish to turn first to Deleuze's argument about the impossibility of philosophy having a "true beginning," as it relies already on the "common sense" about what it means to think: "It is because everybody naturally thinks that everybody is supposed to know implicitly what it means to think."[24] Philosophy can, he writes, only claim to begin without presuppositions, for it does not acknowledge that what precedes it is precisely a dogmatic image of thought that preconditions what it means to think. Throughout his work, and later also in collaboration with Guattari, Deleuze argues that this dogmatic and moralistic image of thought prevents actual thought from coming into being. Thinking must therefore break free from it. In *Difference and Repetition*, Deleuze sets out to identify the postulates of this image, which stipulates what thought is and how thinking operates. Most importantly, thought based on this image—whether in view of Plato's *Theaetetus*, Descartes's *Meditations*, or Kant's *Critique of Pure Reason*—is deeply committed to the "model of recognition."[25] The image is necessarily representational because it accepts the subjective preposition of a universal cognitive faculty that speaks for—or in the place of—others. Hence, thought that relies on recognition and re-presentation will, according to Deleuze, necessarily reproduce this dominating image of thought and its laws: the unquestioned common sense according to which everyone knows what it means to think, the belief that thought has a natural relation to truth, the assumption of the good will of the thinker, and ultimately the superiority of the thinking subject. The image of thought developed pervasively alongside Western philosophy and culture and thus provided a fertile ground for the inception of Eurocentric concepts such as reason, rationality, the subject, the contract, consensus, and justice, as well as their concomitant assumptions about scientific and moral progress, speciesism, anthropocentrism, colonialism, and, ultimately, the hegemony of European intellectual history. These strongly political implications are even further foregrounded in *A Thousand Plateaus*, where Deleuze and Guattari together take up the issue of the image of thought that covers all of thought, arguing that thought is "already in conformity with a model that it borrows from the State apparatus."[26] Here, the image is described as possessing two heads, one being "the imperium of true thinking operating by magical capture, seizure or binding, constituting the efficacy of a foundation (*mythos*)," and the other "a republic of free spirits proceeding by pact or contract, constituting a legislative and juridical organization, carrying the sanction of a ground (*logos*)."[27] What thought gains from this is "a gravity it would never have on its own, a center that makes everything . . . appear to exist by its own efficacy or on its own sanction."[28] Put differently, representational thought does not simply become the law that creates the world in its image. It is, in fact, inextricable from power and legitimizes the thinking subject as sovereign who perceives the world as a mass of individual bodies and objects—to become property—and a blank map—to be territorialized. This (mode of) thought—State, arborescent, or binary thought—is a proprietary and colonizing one. It seizes, captures, and arrests, and it does so over and over again.

In his earlier works, Deleuze emphasizes the necessity of destroying this image of thought to make other forms of thinking—thinking without repeating its own presuppositions—possible. Later, and especially in the course of his collaboration with Guattari, the plane of immanence—the "Being-thought" and "Nature-thought"—arises as an alternative "image of thought," which is neither an image nor representational.[29] In other words, in order to challenge a mode of thinking that only repeats and reinstitutes what has already been established, thinking has to think the "the image thought gives itself of what it means to think."[30] It has to sense and make sense with (rather than of) difference, thinking-with the very potentiality of what is, can be, might be, and would have been in a way that cannot be grasped by representation, analogy, comparison, or recognition. What this makes possible are modes of not only thinking and doing theory differently but of *difference doing theory*.

Precisely because critique—as proposed by Kant or Hegel—is not radical enough to counter this image, Deleuze and Guattari propose a rhizomatic model of thought (namely *nomad thought* or *counterthought*) that aims to destroy it. Counterthoughts are thoughts without image, violent (in terms of force(s) and strength) in their acts, discontinuous in their appearances, and placed in an "immediate relation with the outside" (i.e., outside of the image of thought and its reproductions) rather than originating in the thinking subject.[31] The exteriority of thought is not absolute; it does not denote an alternative image in opposition to the dominant image but rather a "force that destroys both the image and its copies, the model and its reproductions, every possibility of subordinating thought to a model of the True, the Just, or the Right (Cartesian truth, Kantian just, Hegelian right, etc.)."[32] "Radical critique" in the form of counterthought or nomad thought is neither reflective nor dialectical but completely destructive of the image of thought and thereby inventive of other modes of thinking. For Deleuze, to think is "to create—there is no other creation—but to create is first of all to engender 'thinking' in thought."[33] Already in his early book on Nietzsche, Deleuze argues that thinking is "never the natural exercise of a faculty" but "depends on forces which take hold of thought."[34] Not only does thought require an encounter with what is unrecognizable and unknown, yet nevertheless sensible, "thought also never thinks alone and by itself."[35] Thought thus always already thinks-*with*. The prefix *counter-* does not refer to a Newtonian idea of force but ought to be read in relation to the moralistic image of thought. These early articulations of what (counter)thought is and how it operates reveal their synaesethical dimension when read together with Whitehead, and also in the collaborative philosophical work of Deleuze and Guattari. This dimension opens up thinking to more than just human minds and brains (or alleged digital simulacra thereof). It allows for an onto-epistemological reconceptualization of thinking as a relational process and practice that operates in neither binary nor dialectic logic, and it is further aligned with Barad's notion of material-discursive practices through which ontic and semantic boundaries are constituted.[36]

Deleuze and Guattari's conception of counterthought relies on the already-mentioned plane of immanence, which is what the authors imagine as a preconceptual, nonrepresentational "image" of thought. Previously, by asking what the plane of immanence can do in terms of facilitating thought's meaningful encounters with nonrepresentational language, we approached the plane as a *matterphorical expression*, aiming to express a dynamism that knows no mind/matter dualism. Here, in asking what thought *is* and how it can do away with the moralistic and representational image of thought, the plane of immanence also operates as a matterphorical image (or field) of thought. In *What is Philosophy?*, Deleuze and Guattari state that thought and nature, the two sides of the plane of immanence, are interwoven in and as movements, which importantly should not be construed as "spatiotemporal coordinates that define the successive positions of a moving object and the fixed reference points in relation to which these positions vary."[37] Thinking and being cannot be separated, they argue, because "movement is not the image of thought without being also the substance of being."[38] To articulate what movement is, Deleuze and Guattari think-with Spinoza's philosophy of immanence, yet they also depart from the physical assumptions underlying Spinoza's *Ethics*. The dynamism sensed by them is, as argued, closer to what Barad aims to express about the void and its dynamism; namely, a "desiring to be/come, a dynamism of in/determinacy: a liveliness, a life force, a yearning towards existence/existing. It is a potential movement or moveability, a potential e/motion."[39] Different laws of *being* moved. Immanently immanent.

Immanence, Deleuze and Guattari remark, makes thinking increasingly difficult precisely because it departs not only from the dogmatic image of thought but also from transcendence: the thinking subject as the origin of thought and agent of thinking and thought as that which not only takes place but also takes an object. Not only are there neither subjects nor objects in thought, Deleuze and Guattari also argue that thought does not constitute a will to truth but instead "a simple 'possibility' of thinking without yet defining a thinker 'capable' of it and able to say 'I.'"[40] That thought is, to a certain degree, *impersonal*, or at least precedes the personal, is also what Whitehead articulates in *Modes of Thought*:

> Thought is the outcome of its own concurrent activities; and having thus arrived upon the scene, it modifies and adapts them. The notion of pure thought in abstraction from all expression is a figment of the learned world. A thought is a tremendous mode of excitement. Like a stone thrown into a pond it disturbs the whole surface of our being. But this image is inadequate. For we should conceive the ripples as effective in the creation of the plunge of the stone into the water. The ripples release the thought, and the thought augments and distorts the ripples. In order to understand the essence of thought we must study its relation to the ripples amid which it emerges.[41]

Thought emerges from within and from the middle. The phrase "like a stone" introduces the possibility of both kinds of *like*, yet the next sentence tells us that we cannot rely on analogy, comparison, or metaphor, as this "image is inadequate." "Like" here does not link what we assume it would; it breaks with the temporality of our modes of thinking as well as with our learned sense of causality and relationality. Once we depart from the image of thought upheld by most Western institutions and canons, we might find ourselves not knowing what the specific temporality, let alone causality, of thought is. We might learn what various non-Western philosophies and modes of thought have suggested all along in various different ways: that thinking does not necessarily have to take (a) place, a subject, or an object but is inherently collaborative and non-proprietary. One never thinks alone, and it is not only the human or only brain cells that think. Deleuze and Guattari offer us yet another example: "The plant contemplates by contracting the elements from which it originates—light, carbon, and the salts—and it fills itself with colors and odors that in each case qualify its variety, its composition: it is sensation in itself. It is as if flowers smell themselves by smelling what composes them, first attempts of vision or of sense of smell, before being perceived or even smelled by an agent with a nervous system and a brain."[42]

Here, too, being and thinking are inextricable, making-sense together (*syn-*). Light contemplates, photons contemplate, force-carrying particles contemplate together. Such a mode of thinking—of being-thought, ontological thought, and *sense*making experiments—cannot be grasped by holding on to traditional notions of scale, causality, space, and time. But this does not mean that thought needs to become bigger, more encompassing, traveling even faster through space and time so that eventually the mind will be able incorporate the universe (as transhumanists, for example, aim to argue). As Barad cautions, "animate and (so-called) inanimate creatures do not merely embody mathematical theories; they do mathematics," and "life, whether organic or inorganic, animate or inanimate, is not an unfolding algorithm."[43] Reason is an expansionist concept, while ontological thought experiments acknowledge that no thought can *be* and no being can *think* in the presence of the ever-expanding and colonizing mind. Not only do electrons, molecules, brittle stars, plants, and asteroids "stray from their calculable paths," they are engaged in "making leaps here and there, or rather, making here and there from leaps, shifting familiarly patterned practices, testing the waters of what might yet be/have been/could still have been, doing thought experiments with their very being."[44]

What is contemplated, what is at stake, what is thought-with, is precisely the potentiality of being-thought. This has nothing to do with analogy or comparison; nothing contemplates *like* the plant or *like* a singular or specific plant. Thought evolves from ?-*being*, from difference(s) as modes of existence, lived relationalities, collaborative engagements, uncalculated togethernesses, and entanglements that do not follow from, but can always only precede, fixed and given entities. Thought *is* becoming-together-apart. A becoming, Deleuze and Guattari write, is always in the middle: "neither one nor two, nor the relation of the two," but the in-between or the block that creates something different. It is always double, generating both a turning toward and a turning away, a moving forward together and departing from each other—in Glissant's words, giving-on-and-with (*donner-avec*). One never knows in advance what an encounter will bring, yet the impossibility of absolute knowledge, recognizability, and predictability does not constitute a threat to or, even worse, an attack on what is familiar and known. One does not think, Deleuze and Guattari write, "without becoming something else, something that does not think—an animal, a molecule, a particle—and that comes back to thought and revives it."[45] The ripples release the thought, and the thought augments and distorts the ripples. The plant contemplates salt, and so both salt and plant become something else. Salt, in turn and by turning, contemplates too. This is also why there is no natural affinity between truth and thought. "If 'turning toward' is the movement of thought toward truth, how could truth not also turn toward thought? And how could truth itself not turn away from thought when thought turns away from it," Deleuze and Guattari ask provocatively.[46] What this suggests is that it is only by means of its encounters— evoking excitement and forcing something to think—that thought takes on its particular tendency. Encounters, we ought not to forget, are not confrontations or meetings between two bounded entities pushing against each other. If considered from the perspective of an onto-epistemology of difference, where otherness is an entangled relation of difference, encountering means becoming, being-thoughts becoming (with) being-thoughts, sensing and making-sense through incalculable und unrepresentable togetherness. Therefore, a *synaesethics* of thought does not ask *who* it is that thinks (as if thinking can be reduced to an entity) and does not assume predetermined relationalities but rather focuses on what thought can do and how it can encounter without seizing, appropriating, and oppressing that which it encounters. Sensemaking *is* a collaborative practice, for better or worse. This is why we are in this together, yet not the same. Becomings relate becomings, cuts cut together-apart. Entangled sense(s) and mind-blowing difference, calling movingly for an ethics of making-sense-*with*: synaesethics.

# Canonizing: Being Unthinkable

The necessity of learning to think—that is, to engender thinking in thought—is a political, legal, and ethical one. As the quote from the Princeton undergraduate student revealed quite clearly, this necessity cannot be considered apart from institutionalized forms of knowledge production and how those institutions secure their power. The policing, regulating, oppressing, and inferiorizing of alternative modes of thinking are means by which to protect the image of thought because its destruction threatens the very legitimation of the institution's power. Within Western epistemologies and trained by Western thought, we are thus taught to produce recognizable and applicable thoughts and concepts. The tools we are given are representational: We make sense of what is unknown using analogy and metaphor, we construct difference using comparison.[47] We are educated in State thought: "logos, the philosopher-king, the transcendence of the Idea, the interiority of the concept, the republic of minds, the court of reason, the functionaries of thought, man as legislator and subject."[48] As Brian Massumi puts it, this is a mode of thinking that builds walls.[49] It protects what it has established as knowledge, it guards the exclusivity of its modes of production, and it satisfies a sense of familiarity by assuring that whatever gets established by this mode of thinking will remain recognizable to those who define what knowledge is and can be. Difference—which bears no resemblance to the image of thought—cannot evolve for, as Deleuze suggests, it "calls forth forces in thought which are not the forces of recognition" but rather "the powers of a completely other model, from an unrecognised and unrecognisable *terra incognita*."[50] Thinking and knowing this *terra incognita*, holding the potential for the perpetual "creation of a future new earth," is an ongoing process of becoming, of thinking-*with* and of being thought-*with*. It is not a matter of discovery. Encounter does not naturally imply conquest. One does not think without becoming something else, for becoming is always double. Truth is turning, and so is the earth. This *terra incognita*, "adsorbed" by the plane of immanence, is a matterphorical expression too. Being(-)thought is a *matter* of (non)existence as well as (non)life, which in turn are inseparable from questions of knowledge and power. In a passage reminiscent of Whitehead's immanent anarchic reason as a mode of thinking from *within* a particular space-time-matter composition, Deleuze and Guattari argue that "possibilities of life or modes of existence can be invented only on a plane of immanence" on which the law of thinking and its juridical features, particularly the legislation of knowledge, can neither take hold nor place anymore. Should not, they ask in an attempt to imagine an ethics of lived relationality, "judge and innocent merge into each other"; should not "beings be judged *from within*—not at all in the name of the Law or of Values or even by virtue of their conscience but by the purely immanent criteria of their existence"?[51]

That being said, it is crucial to clarify that the opening up of the possibility of nonrepresentational thought, despite the important ethical implications of the destruction of the dogmatic image, counterthought, or any thought for that matter, is not per se nonviolent, let alone inherently good—not even better for that matter. It simply means that the particular violence inscribed in the particular law of thinking that is preconditioned by the image of thought (structuring Western philosophy and, relatedly, also its underlying notions of ontology) will reproduce itself until it is destroyed and new ways of thinking are opened up instead. For some lives and modes of existence, the violence of the image is productive and protective, as it confirms their position of power. For others (be they humans or nonhumans), it colonizes their thoughts and their land, exploits their bodies, renders them inferior, and declares their worlds unthinkable. Counterthought (or nomad thought) is "primary trespass and violence" because it requires breaking free from and destroying an image that covers all thought and renders alternatives unthinkable.[52] Breaking open the moralistic image of thought, which insists on the good and true nature of thought and the good will of the thinker, requires an ethics of thought prepared to avoid and address its own hitherto unknown forms of violence. However, the fact that the potential danger might yet be unknown does not justify upholding a violent structure simply because its violence is already known to those for whom it is beneficial, or at least bearable. In fact, the relation between what is known and what is not, the different ways in which thinking encounters the unknown, and the potential that becoming unknown carries for opening up different modes of existence (and, as Deleuze and Guattari write, an earth or people to come) all invoke different, noncomparable forms of violence. What becomes clear, however, is that wherever and whenever the questions of what is thinkable, what can be thought, and who or what is allowed to think it arise, there lurks not only the question of existence but also of life and livability.

In her 1979 novella *Kein Ort. Nirgends* (No place. Nowhere), Christa Wolf, an East German writer, gives expression to a solitude that results from thinking against the image of thought and remaining unknown and unrecognizable.[53] It leads to a life that is unlivable in the present but remains indicative of an inhabitable world to come. Counterthought, being a traitor, and thinking-with more-than-friends are far from common sense and, because they cannot relate to the common, attest to an "absolute solitude." Yet, Deleuze and Guattari remark, it is "an extremely populous solitude" that is "already intertwined with a people to come, one that invokes and awaits that people, existing only through it, though it is not yet here."[54] Wolf's novella stages an impossible encounter between two young German literary figures, Karoline von Günderrode and Heinrich von Kleist, who both, independently from each other, committed suicide in 1806 and 1811, respectively. The novella—in its style, its attempts to express, and the encounters it enables—is an artful plea for the destruction of the image of thought that underlies the German literary canon. In classical German literary history, Günderrode and Kleist, who (as far as it is known) never actually met, are rendered outlaws, foreign and unknown to hegemonic German thought: Günderrode writing in a time when women were not supposed to participate in the cultural production of an evolving nation's imaginary; Kleist unwriting a time he had long outwritten. "What is unsolvable has no form yet"; what remains is an "unlivable life. No place, nowhere."[55] It is thus only through Wolf's text, and in her own attempts to break classical German thought, that these counterthoughts can take and give (each other a) place. In the text, the two figures finally encounter each other at a salon party in a town on the Rhine, surrounded by, among other renowned party guests, the German legal scholar Friedrich Carl von Savigny, his wife, the poet Clemens Brentano, and his sister Bettina Brentano. Günderrode and Kleist leave the party together; they don't fit, they are uncanny, unrecognizable, foreign. They are even unknown and foreign to each other, but close in this unknownness and foreignness.[56] The encounter—when Günderrode-thought and Kleist-thought meet—reveals the populous solitude that is intertwined with a people yet to come: "To understand that we are a draft—perhaps to be discarded, perhaps to be picked up again"—in any case "referenced to a work that remains open like a wound," understood by beings (*Wesen*) "that have not been born yet."[57] If thought searches, it is less by means of a method than by making uncoordinated leaps.[58] It involves trespass and suffering; it involves becoming (with) what is unknown. "Oh, this innate bad habit of always being at places where I don't live, or of inhabiting a time that has already passed or hasn't yet come."[59] It is a law (*Gesetz*) that "time misjudges us," and yet "time seems to be trying to bring about a new order of things, and we will not experience (*erleben*) anything but the overthrow of the old order."[60] The absolute solitude reveals the lack of any acceptance of yet-unthinkable thoughts or "insane diagrams" (*irrsinnige Pläne*) but also the imperative for those thoughts to be thought—despite the fact that, or perhaps precisely because, (a) life's livability is at stake: "We are lonely. Insane diagrams which send us onto that eccentric track . . . Even if we are prepared to die, the wounds humans have to inflict on us nevertheless hurt; the pressure of the iron plates as they come closer to crush us or urge us to the very edge, gradually takes our breath away; short of breath, full of dread, we must go on speaking, that we know."[61]

The rejection of what has been established as knowledge, the threat that the unknown and the unrecognizable pose for common sense, the readiness to use the utmost violence to squash counterthought on the one hand and the force of counterthought carried only by its necessity to persist on the other: All expose the complexity and precarity of a thinking that has to take (its) place *in* thought. The power of the yet-unrecognizable *terra incognita* forces those who sense an unthinkable yet possible future (as well as past and present) to think a people and an earth to come. Günderrode and Kleist, and certainly also Wolf herself, sense it: "We remain unknowable to us, unapproachable, addicted to disguise. The names of strangers we get for ourselves. The cry of lament forced back into the throat. Grieving is forbidden, for what losses have we suffered? I am not I. You are not you. Who is 'we'?"[62]

What is also articulated here, especially in the proclamation of an "unlivable life," is not only the relation between *being* unthinkable—either unable to think or impossible to be thought of—and *being* ungrievable. It is further the relation between an unthinkable and, to think-with Butler, an ungrievable *life*. As Butler defines it, an "ungrievable life is one that cannot be mourned because it has never lived, that is, it has never counted as life at all."[63] In terms of modernist law and how it defines life—as I will touch upon later in the last part—the link between being rendered unable to think (that is, to think rationally) and the right to (a) life is one of law's defining features. Brain functions, self-consciousness, and the mastering of (human) language—i.e., the possession of reason and the intactness of its associated organ—are prerequisites for having a right to (a) life. What's more, certain lives and modes of existence remain unthinkable precisely because the framework in which thinking and knowledge production can take (a) place cannot accommodate any divergence that would position them in the realm of the thinkable.[64] Not only are modes of being and knowing inseparable but so are those of living and thinking.

In *Poetics of Relation*, Glissant casts light from yet another perspective—from an abyss—on the relation between what is unknown, unthinkable, and unrecognizable and the suffering that accompanies the destruction of the image of thought that colonizes land, builds slave ships, ends lives, deports bodies, and renders difference both unthinkable and thoughtless. The text opens with an attempt to put into words the abysses of the "absolute unknown" that have inflicted unspeakable pain, suffering, and violence onto slaves forcibly taken from Africa and brought them to an utterly unknown land, language, and image of thought. Traveling in one direction, the unknown is chased by determined conquerors and eager discoverers. It is approached by a will to com-prehend, understood in its etymological sense of "grasping" and "seizing," which Glissant calls out. Shores mark the beginning of a new colonizing adventure, legitimized by old thoughts. What becomes known to the conqueror is that which can be captured and arrested, mastered, possessed, territorialized, and shipped across the ocean. Ships sailing in the other direction, however, force abducted bodies into an unknown—abyss after abyss—that is pregnant with lethal spaces and burial grounds. Shores threaten to further silence arriving modes of thought and mark an unknown territory, initially even impossible to think: "But for these shores to take shape, even before they could be contemplated, before they were yet visible, what sufferings came from the unknown!"[65] Indeed, those who survived the crossing of the ocean faced an unknown land—with its unknown sounds, animals, people, laws, smells, histories, landscapes, and languages—and were forced to think-with it in order to survive. "The woes of the landscape have invaded speech, rekindling the woes of humanities, in order to conceive of it. Can we bear ad infinitum this rambling on of knowledge? Can we get our minds off it?"[66]

Here, neither livability nor modes of existence and survival can be untied from the violent potential of both what is known and what is unknown. This is how, as Glissant states, "the absolute unknown . . . in the end became knowledge."[67] This passage from the unknown to the known is not marked by comprehension, that is, by capturing and possessing knowledge, but can instead be understood as becoming, or, in Glissant's terms, as giving-on-and-with. In other words, rather than always declaring and appropriating an object that is thereby rendered "known," the relation between known and unknown is nonhierarchical, and their encounter not only changes but reinvents both. What becomes knowledge is not the *same* knowledge nor even the same kind of knowledge that enabled building the ship or rendering slavery legitimate, but one that relies on a different mode of thought altogether: "the thought of errantry—the thought of that which relates" rather than separates, seizes, appropriates, or draws lines on maps.[68] The one who is errant, "who is no longer traveler, discoverer, or conqueror," Glissant writes, strives to know the world yet knows that it is never completely knowable—not graspable—to one knower. And thus the one who is errant also knows that this unknowability "is precisely where the threatened beauty of the world resides."[69] Errant thought then is a nonproprietary encounter with the unknown and, as Deleuze maintains, with what is unrecognizable—not yet thinkable but sensible. "Something in the world forces us to think" that is not an object of recognition but a fundamental encounter that can only be sensed. As per Deleuze, that which forces us to think is *imperceptible*, as far as perception relates to our (human) sense(s) and as far as perceptibility is tied to recognizability. Yet it is still sensible because it is *in* and *of* the world. This encounter does, and requires doing, "violence to thought" and to the image, as it forces thought to think not according to its principles but according to the exteriority of its relations. One must make relations "the hallucination point of thought, an experimentation which does violence to thought."[70] If there is violence in Glissant's concept of Relation—a particular concept of immanence and relationality—it is in the sense that it destroys the inherently violent image of thought that continues to generalize and colonize. The "most peaceful thought is," he states, "in its turn a violence, when it imagines the risky processes of Relation yet nonetheless avoids the always comfortable trap of generalization. This antiviolence violence is no trivial thing; it is opening and creation."[71] The people to come, understood in the Deleuze-Guattarian sense, inhabit an immanent world of Relation, teeming with differences that *matter* and that are neither binary nor oppositional. Decolonization, Glissant writes, "will have done its real work when it goes beyond that limit," namely the limit of the oppositional and of alterity. This world to come is one where an ethics of difference, rather than of universal principles and generalizations, is called for. It was the Western world, as Achille Mbembe states, that considered itself "the center of the earth and the birthplace of reason, universal life, and the truth of humanity," and it was the West alone "that had invented the 'rights of the people,'" an exclusive concept that only slowly includes those who were not thought of when the highly selective meanings of "rights" and of "people" were defined.[72] As Glissant cautions, "thought of the Other"—the recognition of alterity—does not preclude sovereign power and is "sterile" without the "other of Thought," which is "always set in motion by its confluences as a whole, in which each is changed by and changes the other."[73] This again articulates that any encounter—even between two unknowns—is a becoming (rather than a call for either inclusion or exclusion) and as such changes those that encounter. "We know ourselves as part and as crowd, in an unknown that does not terrify," Glissant writes. This is perhaps also the celebration of the "possibility of black thought" and of "*life* as black thought" that Moten describes.[74]

But how do we think and express nonrepresentational thought? How can we imagine and express, within highly regulated academic disciplines that require representational tools, a mode of thinking that does not (as Glissant states so eloquently when opening his *Poetics of Relation*) withdraw "into a dimensionless place in which the idea of thought alone persists" and that also "matters-*forth*"?[75] How do we encounter the unknown and unthought without forcing it into the molds of familiarity and recognizability? How do we think what is neither subject nor object, what exists in and as non-oppositional difference, and what thus cannot be approached by means of analogy, comparison, or sameness? How do we think what precedes the construction of concepts, and how can we express it in language while ensuring that language neither reintroduces the image nor remains irreducible to representation? And why invest time in asking such questions, writing about a plane of immanence, thinking-with lobster and brittle star, investigating Whitehead's concept of speculative, anarchic reason, reading a novella, and, to spoil the surprise, ending a chapter with a poem?

Because it *matters* which *thoughts think thoughts*. In its most extreme, the stakes of the commitment to a different mode of thinking (which necessarily entails the destruction of the image of thought) range from the decision to follow or oppose particular philosophical predilections to countering imminent threats to both the livability of life and the thinkability of modes of existence. Between reason and the unthinkable all kinds of bodies are hurting, starving, drowning, bleeding, freezing, burning, shaking, falling, rotting, disappearing; all kinds of modes of existence are made impossible. The hegemony of the image, as well as the adamancy with which we are trained to think according to it, fosters and upholds the political, legal, and moral systems that continue to colonize territories and that are inextricable from the modes of oppression accompanying and enforcing them. Deleuze is aware of that fact when he, in *Difference and Repetition*, asks critically whether difference has to be "mediated" in order to "render it both livable and thinkable."[76] Dissolving the mediation, the image of thought, and the modes of thinking based on it and learning not only to think differently but to think difference differently and onto-epistemologically are prerequisites for an ethics of (entangled) modes of existence. They carry the potential to make livable lives thinkable and thinkable thoughts livable. As Mbembe writes in the epilogue to *Critique of Black Reason*, thinking through possible futures, through what must come, is necessarily "*a thinking through life*" and "*a thinking in circulation, a thinking of crossings, a world-thinking*."[77]

Rethinking thought and what it *means* to think almost inevitably touches, as a result of the dominant history of thinking as sensemaking, upon questions of critique. Thoughts that matter for more-than-friends are not to be found in a transcendent realm and its courts of critique precisely because critique should not be understood as the criticism of *false* contents but, Deleuze argues, of *true* forms.[78] Thus, critique ought to create a different, nonrepresentational image of thought. Or, as Barad expresses it, critique has to be understood in the sense of an "immanent critique"; that is, as "*a way of engaging with what matters by taking account of the fact that each entangled phenomenon is already inside of the other*" and never "letting the foundations of thinking sediment to the point where it prevents us from thinking thoughts we need to think."[79] To think matterphorically then is to think critically and creatively *with* what thought, entangled with the world, senses, rather than thinking *about* something as the object of critical thought, judging from an outside. The fact that Deleuze ends the chapter on the image of thought by asking not only what a thought without a moralistic image is but also how it operates "in the world" (rather than *on* the world) already foreshadows the importance of an ethics of thought that is inseparable from both the world and its modes of existence.[80] It is both a *synaesethics of matterphorical thought*, refusing to understand thinking as a representational and transcendent human practice, and an *onto-epistemology of difference* guided by anarchic reason, becoming, and differencing-entangling that bears the potential for thought to *matter* differently—understood not only as operating in the world but as being inseparable from it.[81] This is why thinking and doing theory *matterphorically* also mean refusing to accept the taken-for-granted equation of thinking and reasoning. For reasoning moves differently; eventually, it falls back on the image that gave birth to a philosophy that began without asking what it means to think and which requires a love for beautiful books rather than an ethics of thinking, living, and existing with more-than-friends.

# 5

# Minds Fucking, Making Love

## EXPANSIONIST THEORY

Earth is the womb of the human race. We are currently experiencing the technical, scientific, and social labor pains that indicate that it is time to leave the womb. Even if population is not the problem that many people think it is, many are still attracted to the great spaces out there. Not only is there space, there are massive quantities of resources waiting to be exploited for the purposes of both the spacers and those left on Terra. —THOMAS W. BELL AND MAX T. O'CONNOR (MAX MORE), *Extropy*, issue 1

This boat is a womb, a womb abyss. It generates the clamor of your protests; it also produces all the coming unanimity. Although you are alone in this suffering, you share in the unknown with others whom you have yet to know. This boat is your womb, a matrix, and yet it expels you. This boat: pregnant with as many dead as living under sentence of death. —ÉDOUARD GLISSANT, *Poetics of Relation*

To say that blacks never fully believed in rights is true. Yet it is also true that blacks believed in them so much and so hard that we gave them life where there was none before; we held onto them, put the hope of them into our wombs, mothered them and not the notion of them . . . This was the resurrection of life from ashes four hundred years old. The making of something out of nothing took immense alchemical fire. —PATRICIA J. WILLIAMS, *The Alchemy of Race*

Before transitioning from earth to outer space and then to the deepest points of earth's oceans (as I will do in part II), I wish to demonstrate why becoming sensitive to matterphorics not only is a matter of ethics but also bears significant implications for how we govern and are governed on earth (and even beyond). Having already encountered the normative force underlying the image of thought, which governs what thinking is and can do, it is time to examine more closely the relationship between thought, law, and governance. This relationship is upheld by established institutions, enforced by market dynamics, and driven by centuries-old desires for colonial expansion. What we will encounter is the counterintuitive fact that a mode of thought that too readily leaves behind the earth and its matters may end up determining the rules guiding life still on it. My aim is to highlight the power underlying these modes of thought and to emphasize that if other ways of living and inhabiting this planet are desired, then these desires must be preceded by a rethinking of thought and its long, deeply rooted, and entangled histories.

What will guide this final chapter of the first part of the book is a question posed by Deleuze and Guattari in *What Is Philosophy?*: "What is thought's relationship with the earth?"[1] This question is meant neither rhetorically nor metaphorically but matterphorically. We must think—actually and collaboratively—while acknowledging that the question of whom to think-with is immensely material. A world-thinking—or, as Povinelli has it, an "analytic of existence"[2]—cannot take place in an assumed realm of representation or in a sphere of (human) consciousness, detached from where (a) life and modes of existence matter. In order to understand thought as related to the earth—not to privilege one planet or set a spatial limit to relationality but rather in an attempt to counter notions such as the digital universe, a world as code, or a sphere of (un)consciousness—thinking has to be understood as emerging from within the world, entangled with it and its energies and matter(s).[3] What then does it mean to think against the earth or even without the earth? To approach this question, it is helpful to consider a formative moment in the history of *transhumanist thought*. This moment was engendered by the convergence of extropian dreams and Californian entrepreneurial spirit at the University of Southern California. In the 1990s and 2000s, this convergence rapidly spread not only across Silicon Valley but also to the East Coast of the United States (e.g., the Massachusetts Institute of Technology) and, although differently expressed, to northern Europe (e.g., Oxford University and the University of Warwick).[4] The urgency of thinking-with different modes of thought, and the stakes of doing theory *matterphorically* in relation to the earth and what (and how it) *matters*, will become clearer when examining the neocolonial tendencies inherent in transhumanist thought. Let me clarify, however, that this does not mean that every person who identifies as transhumanist subscribes to neocolonial practices—far from it. It is thus even more important to understand this image of thought and its assumptions about what thinking is—including whether it takes objects, whether it is the faculty of a subject, how thought encounters the unknown, and how this predetermines the production of knowledge—lest the mistakes of the past are reiterated. This also has important implications for contemporary conversations about new digital technologies. If thinking is, for example, assumed to be the faculty and property of a rational, sentient human being, operating through representational thought and detached from the world and its matter, it becomes easy to confuse thinking with computation. This leads to the assumption that artificial intelligence (AI) will one day "outsmart" us all. Large parts of the literature on the dangers and the potential of AI are based on an equation of computation and thinking. The conception of artificial intelligence as the successful evolution and technologically induced perfection of rational thought relies on a notion of transcendence rooted in *humanist thought* that is intensified by *transhumanist thought*. In whatever way we might conceive of thinking, what is certain is that computation is but one mode of sensemaking—and certainly not the most preferable one when it comes to nonbinary relationality that matters differently and for more-than-friends.

Departing from the assumptions guiding the dominant image of thought, which transhumanism intensifies, and attending to immanence through synaesethics is fraught with danger. This is not only because, as Deleuze and Guattari argue, immanence entails all the dangers philosophy already must confront but also because "the problem of immanence is not abstract or merely theoretical"—it "engulfs sages and gods."[5] Put differently, thinking immanence and engaging with matterphorical expressions of immanence challenge the dominant mode of thought built on representation and transcendence. This challenge extends to everything built upon that mode—every word, law, claim to power, appropriation of territory, and conception of knowledge. Consequently, this understanding of theory as an ongoing synaesethic engagement with the world and its matters is not only an ethical imperative. For the unrepresented and unrepresentable—those denied the capacity to transcend what is deemed transcendable—it is a matter of existence, livability, being thought, and being thought-with. In contrast, for those who fear that their power is threatened by a mode of thinking that attends to existential relationality and difference—both gods and sages alike—this way of thinking, which rejects premade entities, dichotomies, and Cartesian dualism, is a call to arms, often literally. This is why, despite the danger, counterthought must break the image and make it as difficult as possible for a colonizing mode of thought—which is not merely ideational but a material-discursive practice—to take (a) place. I argue that (trans)humanism rests on the power and hegemony of this colonizing mode of thought, with its own histories and imagined futures.

The first quotation at the beginning of this chapter, announcing that it is time to leave "the womb of the human race," is taken from the inaugural *Extropy* issue, published in August 1988 by Max T. O'Connor (Max More) and Thomas W. Bell, who, as mentioned, were then two graduate students at the University of Southern California. The printed magazine was published biannually beginning in 1988—from 1989 under the name *The Journal for Transhumanist Thought*—before going online in 1997, when an *Extropy* email list was established, which enabled a wider reach. The striking differences between Glissant's and the *Extropy* editors' use of "womb" in relation to a shifting present and the unknown future of lives, life forms, and modes of existence are no coincidence. The vast contrast between extropian thought, ready to further take place and space, and that articulated by Glissant, the necessity to live and think differently in order to survive this regime, is even stronger when read in relation to Patricia J. Williams's words emphasizing an embodied present and future that carry, literally and matterphorically, the past with them and create life "out of nothing."[6] To add insult to injury, the context for More and Bell's words is their attempt to formulate the pillars of extropian philosophy, which will, they optimistically think, lead to a future where "we will have intelligent machines, material abundance beyond our dreams, vastly expanded intelligences and senses, a pollution free environment, the ability to perfectly simulate any experience, and other wonders still undreamt."[7] The extropians' pronounced interest in science and technology, such as AI, cognitive and neuroscience, life extension technologies, and later cryptotechnology, is guided by their assumed potential to allow extropians to live in accordance with an "unlimited model of reality and self"—becoming "an übermensch," maximizing the potential of human intellect, reason, and consciousness.[8] Extropian desire is directed toward the transcendence of limits and limitations, be they legal, cognitive, psychological, or even physical: "Down with the law of gravity! By what right does it encounter my will?" The presumably "liberating potential" of electronic money, digital cash, spontaneous and decentralized orders, polycentric (private) law, competing private currencies, and idea futures (coupons), as well as artificial selves (minds), immortality, and outer-space colonization, features dominantly in the pages of *Extropy*.[9] In the 2013 publication *The Transhumanist Reader*, More refers to this philosophy as "transhumanism," describing it as a "life philosophy, an intellectual and cultural movement, and an area of study."[10] Its "healthy legacy of the humanist roots" is reflected in "its commitment to scientific method, critical thinking, and openness to revision of beliefs."[11] Transhumanist thought had already been around for some time by 1988, even though it had not yet consolidated into a theory or movement. Yet it was only with the founding of the *Extropy* magazine, with its particular visual language and deliberate use of "memes"; the establishment of the extropy email list; and the creation of the Extropy Institute that transhumanism took on its current characteristics and became more and more known—predominantly among male academics in California, the East Coast of the United States, and northern Europe. Indeed, *Extropy*'s inauguration, and the unfolding, growth, and spread of the ideological goals of its founders and contributors, marks a crucial point in the development and enforcement of transhumanism. Precisely because it matters what thoughts think thoughts, what knowledges know knowledges, what differences differentiate differences, and what relations relate relations, it is crucial to understand which thinkers and philosophies transhumanists think-with. It is necessary to examine how transhumanist relationality is conceived and what ideas of thinking, knowledge production, and sensemaking emerge consequently. Indeed, it is imperative to think about the tradition, image, and modes of transhumanist thought—not only because of transhumanists' refusal to think-with modes of thought unfamiliar to them but also because transhumanist thought proves itself susceptible to colonizing and proprietary endeavors.

For now, and with the differing contexts of the wombs in mind, I wish to return to the *Extropy* magazine and demonstrate how it laid the foundation for the political, legal, and social imaginations conjured up by transhumanism. As indicated before, transhumanism is often confused with various forms of posthumanism, mainly because both transhumanist and posthumanist theories have found traction among scholars, thinkers, and artists. Transhumanism has also gained attention from entrepreneurs with an affinity or curiosity for technological and scientific development. Regardless of interests in fields such as AI, cognitive science, robotics, neurotechnology, nanotechnology, blockchain technology, and biotechnology or suspicions about developments in these areas, transhumanism is not simply "about technology." Rather, it denotes the intensification of precisely those humanist assumptions that seek to justify the superiority of humans not only over nature but over other humans, those deemed incapable of thinking or undesiring of a world in the image of its presumed creators. In the context discussed here, these assumptions merge with another strong narrative: the American frontier, manifest destiny, and anarchocapitalism. It is while considering these characteristics that I am looking at extropianism and, consequently, transhumanism.

In the aforementioned first issue of *Extropy*, the two editors declare their aim to create a futurist movement, a philosophy, and a mode of living and thinking that defies entropy and hails the limitless outward expansion allegedly inherent in humans' nature. The cover shows sixteen rotating "Yang Yin-Yang" symbols, designed by More and arranged in a four-to-four square setting. More writes that the "(Yin) Yin-Yang is an ancient Chinese symbol representing the universe as consisting of co-dependent dualities in eternal flux." Traditionally, More writes, "the Yin-Yang is represented in a circular form"; he terms this the "Yin Yin-Yang" "because its curvilinear form suggests the feminine aspects of the Yin force," which are "passive, intuitive." As a complimentary opposite then, More has created the "Yang Yin-Yang," representing the dominance of the "active, critical, masculine" force.[12] Not only does the visual language leave, intentionally or not, a bitter aftertaste at best, the discriminatory gender inscriptions are plainly evident. The purported complementarity of the "Yang Yin-Yang" force is an attempt to legitimize a (human) force on par with the universe. It makes intentionally apparent the vision that these young men, and some others at University of Southern California, had for the philosophical and ideological future of what would become known as *extropianism* and later merge into what is nowadays known as *transhumanism*: the power and force of a masculine will—and, yes, More has his own interpretation of Nietzsche—to use all available tools to transcend the internal and external limits of human existence in order to dominate, exploit, colonize, and claim property. As More writes: "No mysteries are sacrosanct; the unknown will yield to the intelligent mind. We seek to understand and to master reality up and beyond any currently foreseen limits."[13]

The topics introduced in this issue as the guiding questions and concerns of extropianism range from artificial intelligence, cognitive science, and nanoscience to the rise of cryonics and life-extension technologies. This issue already makes clear that the desire is to enhance humans, to reject any limitations (be they physical, legal, economical, psychological, cognitive, etc.), and to colonize both the earth and the universe. A radical libertarian free-market society and a proactive strategy to influence minds are noted as means to reach their aim of absolute power and transcendence. In terms of the latter, the authors promise

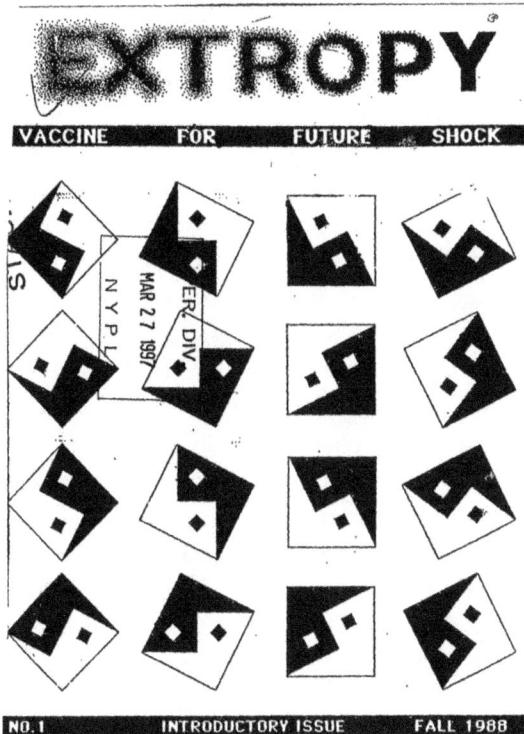

FIGURE 5.1. Cover of *Extropy*, issue 1, Fall 1988, California

to "fuck with your mind and to give you tips on how to make more effective love to other minds" because *Extropy* is interested in exploring ways in which "mindfucking can be used to open up people's brains to extropian perspectives." The goal is even to "go beyond the simply mindfucking to advanced forms of mental sex."[14] As amusing as it might sound, the context is rather disconcerting. In fact, the fucking and penetration of minds in order to spread ideologies and engender the reproduction of these ideas (what a peculiar way to reinforce and reproduce the image of thought!) is further developed into "memetics," the study of "ideas that can replicate and evolve."[15] In referring back to the coinage of the term *meme* by Richard Dawkins in *The Selfish Gene*, More and Bell aim to research memetics in order to "spread the extropian meme itself."[16] Less surprisingly, Dawkins's *The Selfish Gene* rests, in terms of method, mainly on metaphor—including the "colony metaphor," understanding the body as a "colony of genes"—of which he purports to make use by "freely mixing the language of the metaphor with the language of the real thing."[17] The creation of the concept of the meme works along metaphorical lines too. "Just as genes propagate themselves in the gene pool by leaping from body to body," Dawkins writes, "memes propagate themselves in the meme pool by leaping from brain to brain" via a process he calls imitation.[18] Analogous to genetical transmission, memes—which can be ideas, tunes, images, catchphrases, or other attention-attracting concepts—enable cultural transmission and can give rise to another form of evolution. In this context, Dawkins refers to his colleague N. K. Humphrey and a comment he had made in relation to a draft of Dawkins's book, stating that "memes should be regarded as living structures, not just metaphorically but technically." Humphrey further elaborates that the planting of "a fertile meme" in a mind leads to parasitizing of the brain and "turning it into a vehicle for the meme's propagation in just the way that a virus may parasitize the genetic mechanism of a host cell."[19] The methodologically precipitous shift from the metaphorical sphere to actual physical processes is characteristic of transhumanist thought. As the media theorist Gudrun Frommherz argues, the structural analogies between material and viral, bio- and info-, are vital to transhumanism as they reinforce "a fractured, reductionist worldview" that assumes that "all that exists is constructed ground-up, and is a compound of smaller entities that are, above all, observable, measurable and, consequently, controllable."[20] In analyzing memetics as a viable method for transhumanism, Frommherz points out that the comparison between genes and memes is only made possible by faulty metaphorical shifts—paralleling function (gene) with structure (meme) and shifting from replication as imitation (gene) to replication as simulation (meme)—that allow for differences to be overwritten by a similarity in effect. Consequently, memes become autonomous and detached from existing ontologies. They become, in other words, autonomous as artificial creations "free of a relationship to reality and unconstrained by any concerns with truthfulness."[21] This is also where the relation between meme and metaphor lies, since metaphor basically has, Frommherz argues, a "prime memetic function."[22] While poststructuralist uses of language in general, and of metaphor in particular, were mainly aimed at de-essentializing and denaturalizing inscribed hierarchies and dichotomies, they also aimed to establish a linguistic and metaphorical force that, at least at some point, was believed to be strong enough to counter these naturalizations. Derrida's "No Apocalypse" essay, as well as his widely cited "Force of Law" paper, are indicative of that. Transhumanists—although strongly reactive against poststructuralism and postmodernism—not only mobilize memes and the interpretative force and memetic function of metaphors but also, at the same time, claim that they provide a scientific and real grounding. Memetics in the service of transhumanist thought therefore becomes a powerful communication strategy that insinuates scientific grounds for this neomodernist worldview, which serves to veil its ideological motivation.[23]

As is perhaps already obvious, transhumanist thought does not seek to stay with the trouble. Instead, it prepares those committed to that mode of thought to leave the earth and conquer space. Transhumanist thought aims to perfect thinking as computation and data processing because transhumanists believe that the explosion and merger of human and machine intelligence will make it possible to transcend earth and its mortal inhabitants. There is less concern for mortal beings, let alone for the "deathists"—those "in favor of death" by being suspicious about the transhumanist strive for immortality—and therefore also no concern for bodies still embedded in a living environment. As More complains, the "deathists" believe that their death is "really no different from the death of anyone else," which is a perspective only possible when nature, humanity, and living things are understood as a whole. The then twenty-five-year-old More utters his disagreement: "But I am not nature. and I am not all living things. I am ME . . . I am a specific cognitive system."[24] Death, for extropians, is "an enemy which destroys personality, value, [and] information (in the brain)" and with these "all possibility of further growth in personal knowledge, wisdom, power, and experience."[25] The death of other human beings denotes a loss of value for those left behind, yet it remains external to them. But because value is lost regardless when death occurs, the transhumanist's task is "killing death," that is, "destroying the destroyer."[26] When thought together with other extropian and transhumanist goals, such as space colonization and, as I will show later, the dismantling of laws, including human rights, such a language and attitude is problematic to say the least. Where it really gets perverse is when More attests that deathism threatens specifically those who are already immortalists and must be challenged to secure, as per More, "our own survival." As if to crown it all, More even follows up by saying that the more people can be made to "see the light, the more lives will be saved."[27] So there is not much empathy for those who even think about staying with the trouble or those still worried about the heightened vulnerability of poor, oppressed, incarcerated, tortured, poisoned, starved, and otherwise violated bodies and modes of existence— whether human or nonhuman.

If thinking, in transhumanist scenarios, remains in any way embodied, let alone entangled with the biological or nonmachinic world, then it is merely an obstacle for overcoming. The biological body, an undesired conglomerate of matter, is, in the words of More, a "flawed piece of engineering."[28] For transhumanists, the body is a machine and the mind a functional process of the brain that can be re-created in any other suitable matter.[29] For advanced *rational thought* to be possible, the mind, as transhumanist and roboticist Hans Moravec puts it, has to be "freed from the bondage of a mortal body."[30] In an attempt to distinguish his understanding of posthumanism from transhumanism, Cary Wolfe describes the latter as an "intensification of humanism" insofar as the "human" is achieved by escaping or rejecting "the biological, and the evolutionary, but more generally by transcending the bonds of materiality and embodiment altogether."[31] His observation is correct. Transhumanism is, at its core, the futuristic attempt to finally realize the Cartesian mind/body dualism, in order to maximize the possibilities for radical individualism and free-market societies. The term *transhuman* denotes the interim state between the mortal human (Anthropos, Man) and the postbiological, immortal posthuman. Proponents of this line of thinking admit that it will certainly take time before humanity reaches the desired "postbiological condition," wherein posthumans will live in digital eternity as nonbiological existences, as purely computational mind power. Emphasis is thus placed upon perfecting technologies like human enhancement, mind scanning, mind uploading, and whole-brain simulation in order to eventually bring about a new species of beings that are "completely synthetic artificial intelligences" or "enhanced uploads."[32] But once this state is reached, transhumanists argue, there will be no more limits for the advanced human race—the colonialization of space, infinite knowledge, exponential growth, eternal life. This claim is possible not only because of a firm belief in new technologies but also because thinking, in the transhumanist imagination, is tantamount to simple information processing. It is no coincidence that transhumanism developed alongside the rise of computer and information-technology studies, cybernetics, and cognitive science, all of which equate the rational mind to information processing, and thus distills the essence of human existence as a constantly improvable entity. As Katherine Hayles argues, the roots of this line of thought lie as much in ideas of the liberal, autonomous self as in information and computer technologies. This makes it possible to understand intelligence as "a property of the formal manipulation of symbols rather than enaction in the human lifeworld" and to conceive of thinking as disembodied.[33] In other words, *reason* must become digital, the thinking mind must transcend earthly mud and matter, and the Enlightenment—or its updated version—must shine its light deep into the universe.

Importantly, transhumanism rests epistemically and philosophically upon a particular mixture of Enlightenment humanism (with its emphasis on rational thought, progress, reason, and individualism) and human speciesism.

Nick Bostrom, an Oxford University professor, the founder of the Future of Humanity Institute at Oxford, and a prominent representative of British transhumanists, describes transhumanism as an "extension of humanism"—with its emphasis on the centrality of humans and individuals and the promotion of "rational thinking"—that leads to a stronger focus on the possibility of enhancing the human organism through educational, cultural, and also technological means.[34] More, collaborating with Bostrom in realizing the transhumanist project yet speaking from the US American strand of transhumanism, similarly argues that humanism relies "exclusively on educational and cultural refinement to improve human nature" whereas transhumanism aims to "apply technology to overcome limits imposed by our biological and genetic heritage."[35] His "Letter to Mother Nature" develops into a contract between transhumanists (speaking for humanity) and "Nature," ends with More stating that humans "reserve the right to make further amendments collectively and individually," and is signed "Your ambitious human offspring." More declares humanity's transcendence beyond any biological or evolutional ties that "Nature" might have assumed. This cutting of ties with "Nature" consists of seven amendments, including the refusal to "tolerate the tyranny of aging and death" and the employment of computer and biotechnologies to "exceed the perceptual abilities of any other creature," expand humans' working memory, enhance their intelligence, supplement the neocortex with a "metabrain," take charge of genetic programming and "achieve mastery over our biological and neurological processes," bring about "refined emotions," and increasingly integrate "advancing technologies into our selves" in order to finally progress from the transhuman to the posthuman state.[36] Steve Fuller and Veronica Lipinksa—two transhumanists who not only propagate radical capitalism but also openly promote neoeugenics—are quite outspoken about the speciesism that drives transhumanism, describing it as "the indefinite promotion of the qualities that have historically distinguished humans from other creatures," such as the "seemingly endless capacity for self-transcendence, our 'god-like' character."[37] And while More has tamed his language since his early days as publisher of and contributor to the *Extropy* issues, he was not as reticent when he declared that extropians "recognize the unique place of our species, and our opportunity to advance nature's evolution to new peaks," and that "the extropian goal is our own expansion and progress without end."[38] Neither did he see an issue in publishing articles that state that we are "no more than a particular pattern of information, a set of data and processing rules," and that we (those identifying with this definition) will "together . . . cast down the laws of statists, moralists, nature, and logic. Our ultimate goal: the singular, perfect, omnipotent power of God."[39] In a piece cowritten with Bell, More puts it quite frankly: "We want more intelligence, more power, more liberty. We want more life!"[40]

## Transhumanist Normativity

While transhumanism, as a philosophy, seeks to go "well beyond humanism in both its means and ends," transcending the limits of the biological body and seeking to expand (digital) consciousness, its commitment to Enlightenment ideas and ideals has been, from the beginning, guided by a desire not only to establish space colonies but, as More declares, to "search for new forms of governance."[41] While space colonization is used there as a mere thought experiment, a state of nature far enough away (in terms of space and time) to safely articulate radical libertarian and anarchocapitalist desires, Bell's contributions to the journal reveal that such experiments are not just ideational.

One contribution to the 1991/1992 *Extropy* winter issue, entitled "Extropia: A Home for Our Hopes," especially lays the groundwork for Bell's decades-long endeavor to advocate for "privately produced law" and modes of private tech-governance. Extropia, Bell writes in the eighth *Extropy* issue, is the "social realization of extropist principles": the establishment of an "ideal home for Extropians." He offers a detailed plan for the Extropia project: from the establishment of the "Extropy Institute" devoted to research, educating the public, and funding Extropian causes to conceptual parameters for "full-scale field experiments," such as "Free Oceana, a free and sovereign community on Earth's high seas" and a *test ground* for "a space-based community that will liberate us from Earth's grip and prepare us to expand boundlessly into the waiting universe."[42] Extropia, Bell writes, will not be a state but "a society based on real consent" and will "resemble a private corporation." Its "residents"—extropians who will have found "a home for their hopes"—will have "voting rights correlating to ownership of property."[43] While Bell acknowledges that the declaration of sovereignty by a private community will most likely be met with legal, financial, and military pushback, he optimistically emphasizes the advantage Extropians will hold as a result of their skills in information technology and memetic engineering: "*This is a battle we can fight on our own turf—and win.*"[44] For Bell, this is a battle worth the fight; most extropians "won't really feel at home" until they can inhabit "an artificial city floating far above Earth's surface."[45] Until then, Extropia will serve as the ideal earthly test ground for extropian ideas. As a first step, Bell suggests that a loophole in international law might allow extropians to establish a free community on the high seas by "declaring the sovereignty of a ship, floating island, or sea platform."[46] Far from being a "womb abyss," this ship signifies the power to start again, from scratch, without concern for what has been. Bell does not hide the frontierism. In fact, the language and the erasure of the colonial history of North America further expose the ideology behind extropians' attempts to "leave the womb":

> The vast expanses of space offer us the opportunity to make a fresh start—or as many fresh starts as we like. New frontiers have always excited the imagination. America was seen as a chance to experiment with new social orders, new religions, and new ways of living. To some extent, especially on the West Coast, this spirit is alive and flourishing, but such experiments are severely hampered by existing governments which lay claim to every inch of the planet.[47]

In a similar tone, Bell justifies the creation of an extropian "community" by claiming the necessity of protecting extropian hopes and principles: "Self-defense calls on us to imagine a new world," the "social realization of Extropian principles."[48] Aware of the implications of the extropian project, Bell advises to "first of all portray Free Oceana as a benign research project, a 'sociosphere II,'" where the limits of the extropian legal and governmental model can be tested without state intervention.[49] Extropia, Bell writes, will not be a state but "a society based on real consent" and will "resemble a private corporation." Its "residents"—extropians who will have found "a home for their hopes"—will have "voting rights correlating to ownership of property."[50] Similar to Elon Musk's opaquely formulated yet intentionally clear call for "short law" in an interview with Chris Nolan,[51] Bell suggests that "only the barest of legal frameworks" should be established while still making a claim to sovereignty: "We who build Free Oceana will own it. We will thus have every right to establish its laws and exclude those who refuse to abide by them."[52] As More and Bell already claim in 1988, space colonies will be devoted to particular philosophies, with the majority of the space colonizers being libertarians experimenting with "minimal government and free market anarchy."[53] Luckily, the authors state, extropians will not be forced to live where those with different political and religious views will live.[54] The authors remain silent as to potential conflicts between different philosophies, political positions, and religious views, let alone different modes of thought and existence. Like More, Bell has not renounced his early views; in fact, just the opposite. As a part-time law professor at Chapman University—a private university in Orange, California—with a focus on special jurisdiction, copyright, intellectual property, the Third Amendment, and Internet law, Bell has developed his early ideas of "polycentric," nongovernmental, and privatized law even further.[55] In his book *Your*

*Next Government? From the Nation State to Stateless Nations*, published in 2017 by Cambridge University Press, Bell advocates for ULEX, an open-source legal system for "special jurisdictions, ZEDES [Zones for Economic Development and Employment], and other startup communities."[56] Bell introduces ULEX by means of analogies to and praise for Unix, a computer operating system developed in 1969 by Ken Thompson and Dennis Ritchie, professing that "it opened up vast new environments for software to colonize."[57] To make sure there is no misunderstanding about the meaning of "to colonize," Bell adds: "With allowance for poetic license, one might fairly say that what the Founders did for America, Thompson and Ritchie did for applications."[58] For Bell, the "future of law" lies already latent in the history of computer science. The main (and flawed) analogy Bell draws is between computer code and law, stating that everything boils down to a "very simple equation"; namely, that "code = code."[59] Computers and governments, he writes, run on code; to him, this means that his proposal to fashion a legal system after a computer software appears as a "near-inevitability."[60] Although the specifics of ULEX demand a more detailed analysis, I wish to briefly provide a glimpse into the rationale underlying Bell's program, which complements a related project that he, along with his client Elevator City Development, Inc., has been involved in: the establishment of ZEDES in Honduras. Also known as charter cities, ZEDES facilitate "the creation of autonomous privatized city-states designed to exist independently from the legal, administrative and social systems"; in this case, the Honduran state.[61] A neoliberal and neocolonial concept, ZEDES are territorial spaces, quasi-sovereign entities meant to attract national and foreign investors. They have legal personality, are authorized to establish their own laws (including their own independent and autonomous courts), and have both administrative and functional autonomy. According to a report by the National Lawyers Guild (NLG) published in 2013, only 6 of 397 articles of the Honduran constitution would be fully applicable to residents of ZEDES. Fundamental rights, such as the inviolability of a right to life, the right of habeas corpus, and guarantees of human dignity and bodily integrity, are excluded.[62] And, as an analysis conducted by the Center for International Environmental Law (CIEL) reveals, it is small farmers and indigenous communities that will be most affected by the implementation of ZEDES, as their proposed locations are largely in the country's most marginalized areas. The report states that current proposals for port expansion and beach resorts all along the Caribbean north coast of Honduras could impact 24 Garifuna communities, including at the port of Trujillo, which would "require the resettlement of 3,500 people from the indigenous Garifuna community of Puerto Castilla."[63] Felix Valentín, who is a member of the Garifuna and serves as the land, territory, and environment coordinator at the Black Fraternal Organization of Honduras (OFRANEH), formulates the issue most clearly: "The first displacement of the Garifuna was when we were taken as slaves from Africa. The second displacement was when we were deported from [the Caribbean Island of] St. Vincent and brought to what is today the coast of Honduras. The third displacement will be the creation of the 'ZEDEs,' the special development zones, because we have nowhere to go. Where would all the existing Garifuna communities go?"[64]

Jari Dixon, a former member of the National Congress of Honduras and part of the opposition party, adds that the land targeted by ZEDEs represents "the most productive areas, by the sea, along the rivers, even archeological sites." The handing over of land to investors, who will be "owners, lords and kings in these territories," is, per Dixon, what "we're calling 'neo-colonialism.'" Dixon warns that if this "social experiment" is seen as successful, it will become the template for other crisis-stricken countries similar to Honduras.[65] Ultimately, after Xiomara Castro was elected president in 2022, the Honduran Congress unanimously repealed the ZEDE legislation amid growing national opposition and social unrest. This legislative change was driven by widespread concerns over sovereignty, land rights, and the potential displacement of local communities. However, ongoing projects such as the libertarian startup-city Próspera—funded by the venture capital company Pronomos, which was founded by Patri Friedman and has Bell serving as an advisor—serve as meeting places for transhumanists and raise questions about the future governance and legality of existing ZEDEs. Bell has also been involved in the Seasteading Institute (SSI), serving precisely the aims articulated through his concept of Oceana, as well as the Startup Societies Foundation (SSF), which focuses on so-called startup societies/smart cities, private and charter cities, seasteads, microstates, and special economic zones and seeks to "connect, educate and empower small territorial experiments in governance—all over the world."[66] As an anarchocapitalist and extropian project, its aim is to capitalize, quite literally, on the possibilities unfolding from the demise of nation states, increasing mistrust in governments and concentrations of wealth around tech-entrepreneurship. The foundation has chosen its board members wisely—among them, the aforementioned Patri Friedman. Friedman, an active facilitator of competitive government projects, cofounded the Seasteading Institute in 2008 with seed funding from Peter Thiel, cofounder of Confinity—the company that merged with Elon Musk's X .com to become PayPal. Musk, who now owns the platform formerly known as Twitter (rebranded as X), once referred to himself as "Imperator of Mars ☺," a nod to his vision of making humanity a multiplanetary species. According to SSI's website, which describes the institute as a "non-profit think tank," sea-steading means "building floating societies with significant political autonomy," an endeavor made possible because "nearly half the world's surface is unclaimed by any nation-state, and many coastal nations can legislate seasteads in their territorial waters."[67] It further claims that building floating cities serves the pur-ported "great eight moral imperatives"; namely, enriching the poor, curing the sick, feeding the hungry, cleaning the atmosphere, restoring the oceans, living in balance with nature, sustainably powering the world, and ending conflict.[68] While such aims are certainly noble, Thiel's own explanation for funding the institute is more credible, stemming from his dissatisfaction with politics and disbelief in the compatibility of freedom and democracy. In the same online piece, entitled "The Education of a Libertarian," Thiel also argues that the ob-jective for libertarians, including himself, is to "find an escape from politics in all its forms." The "*critical* question," per Thiel, is "how to escape not via politics but beyond it." However, as there are "no truly free places left on our world," such escape is difficult to achieve. Thiel's solution therefore must involve "some sort of new and hitherto untried process that leads us to some undiscovered country." What promise to reach such goals are new technologies capable of creating "a new space of freedom." As examples, Thiel lists cyberspace, which remains virtual; outer space, which cannot yet be inhabited; and, most prom-isingly, seasteading, which he deems technologically and economically feasible. Thiel closes by wishing Friedman the very best in his "extraordinary experi-ment" in propagating the "machinery of freedom that makes the world safe for capitalism."[69] In this context, it should not be surprising that the SSI—with its slogan of "opening humanity's next frontier"—has organized an event entitled "Disrupting Democracy—Creating Zones for Economic Development and Employment in Honduras" in support of the establishment of ZEDEs.[70]

Projects such as the SSF and the SSI may not immediately register as extro-pist or transhumanist endeavors, let alone academic ones, yet the role (and complicity) of academic thought and knowledge production (i.e., that which is situated within institutions of higher education) must not be underesti-mated. Peter Thiel, for example, received his BA in philosophy and his JD from Stanford University, where he also cofounded the prominent conservative and libertarian journal *The Stanford Review* in 1987. Patri Friedman, too, attended Stanford University, in addition to the New York Institute for Technology and Harvey Mudd College—not to speak of the academic affiliations of his father David, who studied theoretical physics at Harvard University (BA) and law at the University of Chicago (JD) and is now professor emeritus at Santa Clara

University School of Law (!), or his much-famed grandfather Milton, who, as mentioned, received the Nobel Prize for Economic Science in 1976. The chairman of the SSF, Mark Frazier, holds a bachelor's degree from Harvard University. The SSF's "research arm," the Competitive Governance Institute (CGI), recently established the *Journal for Special Jurisdictions*, which has published issues on nonterritorial governance (2020) and ZEDEs (2021) and will focus next on "the history, theory, practice, [and] critical analysis" of special jurisdictions in the United States.[71] While Friedman serves as advisor to the CGI, McKinney (also CEO of the SSF) is its codirector and Bell is its academic director; they are also publisher and editor-in-chief, respectively, of the journal. This means that experiments in competitive governance, historically rooted in extropist and transhumanist thought and allied with anarchocapitalist principles, are not merely marginal phenomena. Instead, they are powerful forces operating on the new governance frontier, supported by top-tier universities and entangled with both governmental and market interests. However, it would not be accurate to claim that every person involved in these political experiments identifies as extropian or even agrees with extropian and transhumanist principles. Personal conversations have revealed that participants in these projects, particularly in contemporary blockchain-governance initiatives—from local community members to external visitors and investors—are often unaware of the underlying assumptions and historical context of this mode of thinking or its propensity for expansion. My rationale for outlining the deep influence of extropist and transhumanist thought on contemporary negotiations of governance is thus to caution that thought and ideas are not merely ideational. It matters which thoughts thinks thoughts; for thinking, according to Deleuze, although provoking general indifference, is nevertheless a dangerous exercise. It is only when the dangers become obvious that the indifference, which has been inherent to the enterprise all along, ceases.[72] In the same way that concepts such as sovereignty, rights, or borders are—for better or worse—enabled by humanist modes of thinking, start-up societies, charter cities, and (certain variants of) network states are guided by extropist thought. What mode of thought enables decentralized and community-based rights? Can we collaboratively conceive of different ways of self-governance, not in conflict but through productive and mutually beneficial relationships with governments, industry, and the environment?[73] Could a *matterphorics of law* provide the guidance we need?

For those unthought (of) and rendered unthinkable, there has always been an urgency for modes of thought that do not aim to take (a) place, transcend the earth or even the world, or claim ownership of both what is known and what is unknown, unrecognizable, and unfamiliar. Indeed, there is an ethical and material need for forms of knowledge production that fall outside of the Western imagination and that cannot be acquired through a love for reading beautiful books on beautiful campuses. As I have argued, what is required is (among other forms of engagement) a practice of thinking-with. The urgency stems from what Haraway calls staying with the trouble. This requires making oddkin, rather than drawing analogies and separating according to them, because "we require each other in unexpected collaborations and combinations."[74] It necessitates "learning to be truly present, not as a vanishing pivot between awful or edenic pasts and apocalyptic or salvific futures," but "entwined in myriad unfinished configurations of places, times, matters, meanings."[75] If, as Barad writes, the human subject is the locus neither of knowing nor of ethicality, and if ethics is not about a "right response to a radically exterior/ized other, but about responsibility and accountability for the lively relationalities of becoming of which we are a part," then an ethics of thought has to address the ways in which we are accountable for what thought can do.[76] This is not simply a question of intent, nor about how to best escape what Kant called self-incurred immaturity. Rather, such an ethics would seek to account for thinking as a material-discursive practice: for how certain modes of thought, whether we admit it or not, not only take (a) place in institutions but also seize land, bodies, cultures, and histories and colonize other forms of knowing and existing. Indeed, it concerns a different mode of thinking and doing theory: *matterphorics* as a *synaesethics of thought* that response-ably attends to the way thought encounters what is yet unknown, unrepresentable, unthinkable even, and to how it either reproduces the already known or refuses to do so. Because all life forms—including inanimate forms of liveliness—do theory and because the world, too, theorizes, collaborative research and practices of making-sense-*with* that are not inhospitable to the earth (including human and non-human modes of existence, matter, and forces) are what ultimately promise to create futures, presents, and pasts for more-than-friends and what can cultivate an ethics of thought as lived relationality through an onto-epistemology of difference. It might well be that the space of collaborative research and thought is not yet something that is, evoking once again the Princeton undergraduate student, already known everywhere on campus and that modes of counterthought will remain unloved. In moments of doubt, it should be recalled that, if there is solitude in counterthought, it is still an extremely populated one—something the Tongan poet and academic Konai Helu Thaman helps us to remember with her poem entitled "Thinking":

you say that you think
therefore you are
but thinking belongs
in the depths of the earth
we simply borrow
what we need to know

these islands the sky
the surrounding sea
the trees the birds
and all that are free
the misty rain
the surging river
pools by the blowholes
a hidden flower
have their own thinking

they are different frames
of mind that cannot fit
in a small selfish world.[77]

**II**

# Forces of Law

*A Matterphorical Case Study on Man Falling*

# 6

# Drinking the ~~Kool-Aid~~ Red Bull

## A MATTERPHORICAL CASE STUDY

Our entire picture of the world has to be altered even though the mass changes only by a little bit. This is a very peculiar thing about the philosophy, or the ideas, behind the laws.
—RICHARD FEYNMAN, *The Feynman Lectures in Physics*

Everything is at stake—one should not change the tendencies of gravity and expect to remain the same. And if you wish to remain as an object affected by gravity, then what?
—ELIZABETH POVINELLI, *Geontologies*

Tragedies require a fall, but not every fall is tragic. Height plays a role—in plays and in ballads, on earth and even in outer space. In literature, the height of fall (the *Fallhöhe*), for example—as Arthur Schopenhauer termed the *Ständeklausel* following Charles Batteux—has been a significant device not only for entertainment but also for cathartically resetting the moral and social compass.[1] The *Ständeklausel* itself fell (the fall of the fall, so to speak) in German literature with Gotthold E. Lessing and in English literature with William Shakespeare, already centuries ago. However, fear of heights, or vertigo (*Schwindelgefühl*), as Walter Benjamin writes in the *Origin of German Tragic Drama*, remains a sensibility not only of the cultural connoisseur but of the scholar who, while critically informed, is still more comfortable with the panorama view, God trick, reflection, and representation in positions where even falling *means* rising. The height of thought, the sunniest side of theory, and the most enlightened reason(s) remain insensitive to falls and leaps that are less perceptible, less massive, less palpable. But the question of scale is a tricky one. Gravitational waves—dents in the fabric of space-time that alter distances and proximities between all bodies in the solar system—have been observed to stretch space by one part in $10^{21}$, "making the entire Earth expand and contract by 1/100,000 of a nanometer, about the width of an atomic nucleus."[2] The Enlightenment period in European intellectual history lasted for about a century, give or take a few years, while detection of ancient light suggests the age of the universe to be 13.77 billion years, give or take 40 million. The strong nuclear force is carried (for lack of a better verb) by virtual, massless particles (gluons). Unlike other forces, it grows stronger with distance by interacting with electromagnetic and electrostatic forces, acquiring the potential to unleash amounts of energy unfathomable to the human mind. In such a light, it seems peculiar that concepts and their relationality are considered unaffected by phenomena inseparable from their constitution. Highlighting this peculiarity in thought, Barad asks, in regard to the strong nuclear force, how a force, despite extending only "a mere millionth of a billionth of a meter in length," can not only reach global proportions and destroy cities in a flash but also reconfigure "geopolitical alliances, energy resources, security regimes, and other large-scale features of the planet" while leaving the geometrical notion of nested scale unaffected.[3] What, we might ask, is high and what is low? What is small and what is large? How do we make sense of distance and proximity? Indeed, what does it even mean to be close to the earth or detached from its matter(s)? How close can thought come to the sun, and at what temperature do concepts get sunburned? Icarus fell from the skies into the ocean because he flew too high (what a bitter aftertaste to Daedalus's eureka moment), exposing his wax wings to the sun's heat, which melted them away. Yet it is only reason, Descartes suggests, that can grasp what wax *is*, determining once and for all that the rule of the mind, *Cogito, ergo sum*, is the "safe haven"—or, as the legal scholar Rudolf von Jhering shows, safe *heaven*—for making sense *of*, rather than with, the world.[4]

Descartes's reflections on beeswax are metaphorical and do not align with modern scientific understandings.[5] Far from being a metaphor, beeswax results from a complex material process involving many activities and becomings, such as plant pollination, reproduction, and dances, that suggest different futures in relation to the position of the sun.[6] Specifically, female worker bees secrete thin, lucid scales from glands on the ventral surface of their abdomen when the hive's ambient temperature is raised to around thirty-five degrees Celsius (ninety-five degrees Fahrenheit); these scales are then masticated and used to construct honeycombs. Given the strong presence of wax metaphors in European intellectual history, their "power as conceptual material," as literature scholar Lynn Maxwell puts it, has been extensively traced and interrogated, even taken as a fundamental material that Enlightenment's lucid dreams are made of.[7] Yet the question as to whether power tastes sweet like honey, let alone inquiries to its melting point, despite desires for social and political revolution, global temperature rise, proliferating wildfires, and many forms of nuclear power, has not yet been considered. Not long ago, theory held that trees could burn without their meaning being affected in the slightest. Times have changed (and so has time). Not only do matters seem upside down, matter itself has become mortal.[8] Temptation still tastes, at least from time to time, sweet, like a forbidden fruit or an overly sweetened energy drink. And while the fall might leave a bad aftertaste, at least for those fortunate enough to occupy (quite literally) the sunny side of economic, political, social, and legal orders, it is celebrated as bringing humanity not down but up—to the moon, to Mars, and beyond. In the wake of the *fall for power*, a phrase whose meaning this section shall further elaborate, modes of living on a planet are being aggressively renegotiated. The atmosphere is thinning, temperatures are rising, and breathability is decreasing along lines of racial, socioeconomic, and gendered divides. Yet matters of power have shifted too. It is not enough to shift perspective, to see like a state, to speak of capitalism in the singular, to euphorically follow each call for "democratization," to be "antifascist on the molar level," to "bemoan complicity," to "remain within in the same mode of thought" and simply adjust it "to the reality of things"—in other words, to fall for the image of thought again and again.[9] Indeed, to speak with Deleuze and Guattari, judgments of God are articulations of a lobster, the paths of which seem further and further beyond critical attempts to trace them out.[10] Complicity is cut-together-apart.[11] Consensus, even if productive of a common enemy, is a powerful liberal concept that excludes incommensurabilities and "rest[s] on the injunction not to err."[12] Theory should not have to start from consensus. Yet there might be string figures or even common threads that can be traced back to tense, meticulously woven webs of power. One such thread can be found, despite significant differences in their histories and aims, in the work of feminist, Black, queer, indigenous, poststructuralist, postcolonial, new materialist, and critical thinkers: Man—not to be confused with "men" in general but understood as the humanist subject of rights, the knowing and thinking mind superior to what lacks reason, Anthropos signifying human exceptionalism and radical frontierism—must fall. He must descend from the sky, where He claimed to join, even replace, the Gods, and from the center of knowing and being. Only then can an affirmative onto-epistemology of difference, a sensibility for how different modes of existence can come to *matter*, and nonexclusive knowledge production become thinkable. And yet, despite the continuous call for Man's fall, and even occasional declarations that He has, the gravitational pull of dominant systems of thought, to rephrase Rosi Braidotti, seems almost irresistible.[13] Braidotti warns us that Man, Vitruvian Man, "rises over and over again from his ashes [and] continues to uphold universal standards and to exercise a fatal attraction."[14]

While Braidotti's statement mobilizes metaphorical references, it also invites us to ask what it might mean to resist this fatal attraction. What if we do not take the demand for Man's fall metaphorically but instead inquire *synaesethically* what a fall might mean? How can such a fall, if we are attentive to the material-discursive practices and *matterphorics* it involves, reveal a different mode of doing theory? How can it foster a mode of concepting that becomes response-able and sensitive to the injustices unregistered and unrecognizable by our current modes of analysis? What if we ask what actually happens when the fatal attraction of humanist ideals suddenly faces mass attraction, not only because—as is the case with major events in an era of satellite and internet technologies—the fall breaks online viewing records but also because gravity, the force of mass attraction, pulls mind and matter together? If Man falls, will He free-fall, that is, with negligible resistance? And if so, what is it that allows Him to fall more freely than other bodies? Is there, despite mass attraction, a terminal velocity to His acceleration? Does molecular resistance still matter in considering His fatal attraction? Or do we have to look even closer: at the subatomic—or in-elementary[15]—where gravity might be negligible, where attraction exceeds the dynamics of pushing and pulling, where indeterminacy expresses difference affirmatively? And further, will He land on His feet, on solid ground, or will He encounter different physical states of matter? If Man falls, what falls with Him? Or, put differently, is it Man who falls, individually and bounded through space and time, who creates the conditions for catharsis, or is it the relationality of the fall—the matters and meanings, atmospheric frictions, molecular commitments—that falls for Man? How high and deep must theory, in this case legal theory, go to *matter*—and where do we go from there?

## The Rise—on the Sunny Side to the Moon, and Further

Before attending to the fall, however, it stands to reason to look at the stakes of the *rise* that, at least for legal theory, turn out not too different from those of the fall. Take Elon Musk, who certainly enjoys flying high—although not too high, like Felix Baumgartner, who ended up in the desert on his knees, or like Icarus, who fell from the sky into the ocean. Concerned about humanity's fall, he is determined to rise, at least in terms of wealth and power; he himself does not plan to physically ascend toward the sun or Mars or even too high up in the air anytime soon. In fact, he has given up aviation entirely. After all, as we read in *The Space Barons*, it became risky, and he has to think about his "kids."[16] While actively pushing Mars colonization, Musk admits that "the first journeys to Mars are going to be dangerous," that the "probability of death is quite high," and that he will most likely not be on board the spaceships, as he would like to see his children grow up.[17] Nonetheless, Musk is confident that people will sign up for the mission because they want to be pioneers, much like the settlers of English colonies.[18]

The stakes (and stocks) are high. Humanity will, Musk predicts, either cease to exist or become a digital, multiplanetary species, "ensuring the light of consciousness is not extinguished."[19] As Mars is about 87.17 million kilometers (0.5 astronomical units) further away from the sun than earth is, this light will have to shine even brighter, bringing new challenges to (the) mind. And while it is not yet clear what the right atmosphere for artificial minds will be—as well as the for rights and law, as we will see—it is safe to say that currently Mars is uninhabitable for most known earthly life forms, including humans. Grand visions demand grand analogies; in this case, even ones envisioned to *matter* intraplanetarily and requiring hitherto unprecedented amounts of energy, at least when it comes to planned manmade projects (it is of course debatable what "planned" means in terms of Anthropocene discourse). The idea increasingly popularized by Musk is to create a colonizable—or, ironically, livable—atmosphere by means of "terraforming," defined by the Oxford English Dictionary as the "process of transforming a planet into one sufficiently similar to the earth to support terrestrial life."[20] Musk's suggestion, known better among his roughly 51.8 million X followers as "nuking Mars," involves two small "artificial suns": more specifically, the deliberate detonation of nuclear fusion bombs above Mars's two polar caps,[21] where the most accessible carbon dioxide ($CO_2$) reservoirs on the planet are located, in order to release the greenhouse gases back into the atmosphere, increase atmospheric pressure (to about 1 bar) and temperature (to above 273 kelvin), and thereby warm the planet enough for liquid water to become stable. When Aleksandr Bloshenko, executive director for science and long-term programs of the Russian space agency Roscosmos, responded in 2015 to Musk's plan to "nuke Mars" by stating that more than ten thousand missiles would be required to achieve this vision, Musk tweeted nonchalantly: "No problem."[22] In terms of law, carrying nuclear missiles into outer space might very well become a "problem" for nongovernmental (and governmental) entities, even despite often-made calls for increased adaptability of legal systems to enable, as one legal scholar puts it most optimistically, the "creative ingenuity of the visionaries and daredevils like Musk."[23]

Science, too, exposes "problems," to put it euphemistically. In a detailed paper published in *Nature* in 2018, scientists explain that, taking into account all nonatmospheric $CO_2$ reservoirs or sinks, there simply is "not enough $CO_2$ on Mars in any known, readily accessible reservoir" that, if released into the atmosphere, could produce the desired increase in pressure and temperature.[24] To illustrate the scope of the problem, the same scientists specify in another piece that it would take "the equivalent of a million $CO_2$ icebergs a kilometer across to terraform Mars."[25] And yet it seems that neither the requirement of ten thousand nuclear missiles nor that of a million nonexistent $CO_2$ icebergs on Mars matter as long as ideas sell. "It might make sense to have thousands of solar satellites to warm Mars vs artificial suns (tbd)," Musk tweeted in 2019. The future, in other words, is yet *to be determined*. And this is what matters to Musk and (his) investments. So, why not look at the sunny side of things? After all, if we believe in those who have proven capable of rising to power, become fueled by their energies, follow their dreams, invest in their visions (sometimes with cryptocurrencies), and let technology do the rest, the future—maybe not for the earth but for humanity—will be bright: multiplanetary existence, ever-extending consciousness, planet hopping, whole-brain simulations, the creation of a thick and breathable atmosphere on Mars, and also, ideally, sustainable-yet-ownable energy on planet earth.

Admittedly, Musk's ideas and his decisions on where to put his energies have paid off: On January 14, 2021, *Forbes* announced that Musk, with a net worth of $182.9 billion, was the "wealthiest man on the planet"—a planet that, we learn from a press release by the NASA on the very same day, had just experienced the warmest year on record. For the United States (Musk's country of citizenship and according to him objectively the "greatest country that's ever existed on Earth" and "the greatest *force* for good of any country that's ever been"[26]), this manifested in a record number of climate-driven disasters, including a devastating record of twelve tropical storms and a total of 10.3 million acres burned in wildfires across the west. Despite the far-reaching consequences of these events and the outbreak of the global COVID-19 pandemic, Musk's wealth surged by more than $140 billion in 2020. In early 2021, Musk was also reported to have sold his last four houses in sunny California, which well before the COVID-19 pandemic had been among the states with the largest number of people experiencing homelessness and had the highest rates of homelessness in the United States. In fact, by March 2020, California's homeless population made up more than 26 percent (more than a quarter) of unhoused people in the United States, with the highest percentage fully unsheltered (72 percent) of any US state.[27] Ironically, this was the very same month that Musk heroically announced on X that he would sell "almost all physical possessions" and soon "own no house" (@elonmusk, May 1, 2020, 11:10 a.m. EDT). On the radio show hosted by controversial comedian Joe Rogan in May of the same year, Musk, wearing an "Occupy Mars" t-shirt, rhetorically asked whether "Mars or a house" was more important; he answered his own question by stating that "allocating time to building a house, even if it is a really great house," is simply not a good use of time "relative to developing rockets to get us to Mars and helping solve sustainable energy."[28] Sitting across from a man who has not only risen to inexpressible wealth but claims to be taking the meme-turned-cryptocurrency Dogecoin "to the moon" and facilitating humanity's rise to an interplanetary species, colonizing first Mars and then the universe, and whose Twitter bio says "Technoking of Tesla, Imperator of Mars ☺," Rogan asks Musk in a different conversation from *where* he is taking all *his* energy to keep up with his work. Without doubt, Tesla, SpaceX, Starlink, Neuralink, and SolarCity take, as well as produce, store, and mobilize energy—and power. Where to begin answering Rogan's question? Revealing the entanglement of energies involved, even in the production of a single Tesla car, would require extensive matterphorical case studies: from the company's workforce and the employees' food, fuel, and muscle consumption to the fact that Tesla (as well as Apple, Alphabet, Dell, and Microsoft) is being sued for "knowingly benefiting from and aiding and abetting the cruel and brutal use of young children in Democratic Republic of Congo ('DRC') to mine cobalt" used for Tesla's car batteries and even the electricity usage of X's data servers that make it possible for Musk to publish tweets that frequently affect the rise and fall of crypto, market, and social values, most recently meme stocks.[29] "I am become meme / Destroyer of shorts," Musk tweeted not too long ago (@elonmusk, February 4, 2021, 5:08 a.m. EDT), a phrase referring not only to his power to affect markets with social media activities but also to Oppenheimer's notorious statement, "I am become Death / the destroyer of worlds," recounting his involvement in the detonation of the first atomic bomb in New Mexico in 1945. Decisions on where to put energy and where to take it from cut-together-apart too. It remains to be seen what force his decisions will carry once he is appointed by President-elect Donald Trump, who, shortly after his reelection, announced his intention to appoint Musk as the new head of the Department of Government Efficiency (DOGE).

Not only are bright minds affecting, creating, terraforming, mining, envisioning, technologizing, and digitalizing the future, they are considered to *be* the future. During the aforementioned Joe Rogan show, Musk estimated that 25 years in the future "more of you would be in the cloud than in the body" and claimed that COVID-19, which he jokingly claims to have never heard of, has "taken over the *mind-space* of the world to a degree that is quite shocking." Confidently, Musk asserts his belief that the COVID-19-related mortality rate is actually much lower than the World Health Organization (WHO) has reported and dismissively states that, for him, after having followed pandemic-related developments in China, observing them in the United States is like "watching a movie in English."[30] This insensitivity to the realities of bodies and precarious modes of existence—an insensitivity that in the COVID-19 context is closely tied to government decisions as to whether Tesla plants can reopen—and the collapsing of these embodied existences into modes of representation speak to Musk's transhumanist mode of sensemaking. When he first appeared on Rogan's show in 2018, for example, he spoke with excitement about "the future" and the expansion of consciousness into outer space but also declared to "love humanity," to be "pro-human," and to believe that, even if it sounds corny, "love is the answer." But when Rogan asks him directly about ideas of (and in) the future, Musk's facial expressions change. Initial hesitation mixes with both uncertainty about what the radio host, who seems unsure of how to react to Musk's cryptic language, might think and confidence in the truth and inevitability of what he is going to reveal: "Darwin is not going away . . . that one will be there." It will, he emphasizes, "just be a different arena." In response to Rogan's question whether he means "a digital arena," Musk responds by reiterating, "just a different arena."[31] This is not the only time he has claimed that Darwin's evolutionary (biology) theory will hold when minds are uploaded and the self is multiple and digital. Musk's fear—one that he certainly shares with many tech-entrepreneurs—is to become outcompeted. In yet another interview, advertising his neurotechnology company called Neuralink that specializes in brain-computer interfacing, he reiterates his fear of the destruction of "humanity as a whole" and sets the potential of AI to destroy "civilization" on par with global warming and nuclear bombs by invoking evolutionary theory. He warns that humans cannot compete with AI, as biological intelligence cannot compete with artificial intelligence. Without humans merging with computers and eventually becoming fully digital, AI will destroy humans "the same way that humans destroyed habitats of primates." His explanation, stating that "when *homo sapiens* became much smarter than other primates, it pushed all the other ones into a small habitat" as "they were just in the way," is followed by the claim that the merger of humans with AI will not only save humanity from such a fate but will also bring the "democratization of intelligence."[32] The short interview ends with Musk stating that although the previous year (2017) had been the toughest in his life, as "Tesla faced a severe threat of death," and although Tesla's near-death experience had shaken his belief in humanity, he still is "pro-human." Ironically, when asked if he ever worries about himself, Musk conformingly responds that "no-one should put so many hours into work," "people should not work that hard," and it is "painful" and hurts his brain and his heart.[33] Speaking about the latter, Musk emphasizes in other interviews that he is human, that he has felt depleted, and that he has relied on "eight cans of Diet Coke and several large cups of coffee a day"—and that, when his heart was broken, he had drunk "a couple of Red Bulls," an energy drink that, as we will see, is closely entangled with the rise and fall of men.[34]

It is here that our matterphorical case study begins: here at the intersection of Red Bull—whose slogan promises not only to give you energy but also wings—and the confluence of power and energy, falling and rising. Methodologically, it is important to note that neither Elon Musk nor Felix Baumgartner nor Schiller's diver (whom we will encounter later) are to be considered villains, idols, or representations of good or evil in this (or any other) world. Instead, they serve as narrative entry points into fields of matter and meaning making. Building on matterphorics as a synaesethics of thought as outlined in the first part, this part presents a *matterphorical case study* focusing on law and legal thought, demonstrating how representational modes of thought operate in and through law and also fuel market-based normativities. Given the history of legal thought and its fidelity to representationalism, the claim that law *matters*—that it is entangled with (and not simply affected by or affecting) forces, attractions and repulsions, and waves and particles, as well as bonds and alliances of all kinds—might already sound counterintuitive. Yet in the same way that a word is not simply a word, a concept is not a concept, and meaning is entangled with matter, law cannot simply be considered a written set of rules, a history of court cases, a collection of international treaties, or a guaranteed transaction via smart contracts. When it comes to legal theory, matterphorics can build on crucial work already done on the matter and materiality of law.[35] Thinking-with this work, it pushes further by arguing for a matterphorical approach to law and legal thought. Rather than stopping at materiality, it goes "all the way down," taking seriously the idea that what matter is (or does)—which, following Barad, is not only an open question but an ongoing ontological questioning that articulates and materializes differences in its mattering—must also be acknowledged and addressed by legal thought. My concern is that if we do not consider the onto-epistemological assumptions on which theories are built—such as what matter or force is/does—then how can we change modes of mattering? How can we address the ways in which some modes of existence come to matter while others are excluded from mattering?

Entry points could have been chosen differently. As the previous part demonstrated, the representationalist image of thought underlies both humanist and transhumanist thought. Musk's investments in humanity's becoming mind, aiming to "shine the light of consciousness into the universe," or John Perry Barlow's 1996 assertions in "A Declaration of the Independence of Cyberspace," envisioning a "civilization of the Mind" as a new form of cyberspace governance, are not rare exceptions.[36] Instead, they are part of how dominant representational modes of thought, intertwined with various economic and political agendas, are imagining and constructing possible futures in space and on earth. While the previous section cautioned against representational modes of thought that deny matter its existence, this matterphorical case study demonstrates that this denial does not necessarily render the body, as the mind's other, insignificant. Rather, the differential mattering of bodies and the modes of existence made possible or impossible are all closely related to these modes of thought, including legal thought. Some bodies, it turns out, fall even more freely. To ask the questions subsequently raised about the force(s) and matter(s) of law and where the potential to shift its mode of un/mattering lies becomes an ethical imperative. In what follows, it will be precisely the fall as *matterphorical expression* and *synaesethic sensorium* that reveals matters of law and force, making sensible exactly where legal thought ought to think-with the matters at stake.

7

# *I Am* Free, Free Falling

It felt like I was in prison for five years. On the very last day when I landed and I finally got out of the suit, it felt like the prison doors opened and I can just walk away. That was a huge relief. —FELIX BAUMGARTNER, Austrian parachutist

I was alone with my crew and the people I had rescued, facing an imaginary wall, built by the governments, brick after brick, in the middle of the sea, where no boundaries are physically possible. —CAROLA RACKETE, Sea-Watch 3 captain, European Parliament speech

Standing at the edge of the platform of a specially engineered balloon gondola, more than twenty-four miles above New Mexico, Felix Baumgartner, an Austrian parachutist, glances down one—a view that is, thanks to special camera and communications systems provided by zero-gravity cameraman Jay Nemeth and his company FlightLine Films, shared with a record eight million YouTube live viewers. Baumgartner—"Born to Fly," as the tattoo on his forearm reads—is determined to break records, limits, and, while he is at it, straight through the sound barrier. Defying walls and limits, crossing boundaries, pushing through resistances, and rushing through environments deemed "hostile" to the human body come with risks; in that regard, Baumgartner's feat is no exception. He is, however, prepared. Red Bull, the Austrian energy-drink company, has spent millions of euros on Red Bull Stratos, the "Mission to the Edge of Space." Baumgartner's suit—a customized "personal life-support system"—was designed to supply the daredevil with oxygen, shield him from extreme temperatures and radiation, and provide his body with all the protection needed to set his feet safely back on earth territory. In other words, all is set: The livestream—although equipped with a built-in delay in case of unforeseen complications and to prevent broadcasting the hero's death live on YouTube—is on. Red Bull CEO Dietrich Mateschitz is confident his company will profit tremendously from Baumgartner's free fall from the stratosphere. "I am going home now" are the words millions of YouTube live viewers hear him speak into his helmet microphone right before they witness, almost from a bird's eye view (a questionable metaphor as there are, of course, no birds in the stratosphere), his fall toward earth.

After roughly nine minutes, four of them in total free fall, Baumgartner's feet touch earthly ground. Then he falls again, onto his knees. The mission earned Red Bull an enormous increase in profit and Baumgartner three world records: With a top speed of 1,357.6 kilometers per hour (843.6 miles per hour)—greater than the speed of sound, which is 1,236 kilometers per hour (768 miles per hour)—he is the first human to break the sound barrier in free fall. He has also completed the highest ever free-fall parachute jump, from an altitude of 38,969.4 meters (127,852 feet), and thereby achieved the fastest ever speed in free fall.[1] The online entry of the Guinness World Records describes Baumgartner's "death-defying, multiple record-breaking leap to Earth" at length before closing with the fact that only two years later Google executive and US citizen Alan Eustace "fell from Earth" from an even higher altitude, thus breaking Baumgartner's record for the highest free-fall parachute jump. The fact that another man has fallen from earth, so to speak, has lessened neither Baumgartner's fame nor Red Bull's meticulously crafted image as the energy provider unwilling to accept limits. In case anyone forgets about Baumgartner's "supersonic free fall," a Red Bull YouTube video freshens up memories, constantly replaying the fall accompanied by the song "Free" by the Scottish alternative rock band Twin Atlantic, produced by Red Bull Records: "Don't be told it can't be done / Because the best all die young."[2]

Baumgartner admits that he was "kind of attracted" to the fact that no one had ever succeeded in breaking through the sound barrier in free fall, although many had tried, and some even died in the attempt.[3] And indeed, in *Stratonauts: Pioneers Venturing into the Stratosphere*, Manfred von Ehrenfried provides a list, starting in the year 1785, of fifty-eight "stratonauts" who were either killed in the stratosphere while trying to fly in it or died shortly after as a result of flying or falling. Causes of death include hypoxia, high-speed impact, crash, decompression, crash landing, loss of control, collision, failed ejection, training crash, and being shot down.[4] The history of stratosphere jumps (or attempts thereof) is guided by not only the will to transcend limits, laws, and physical boundaries but also its failures—although often fatal, no less heroic. In *Magnificent Failure: Free Fall from the Edge of Space*, for example, Craig Ryan purports to tell "the *real* story" of one of Baumgartner's predecessors, Nicholas Piantanida, a New Jersey–born truck driver and parachutist who died in 1966 from injuries suffered during his third attempt to jump from the stratosphere.[5] In his first attempt to break the record for the highest free-fall parachute jump, the balloon ascent failed. His second attempt failed when, after already reaching an altitude of 123,500 feet, Piantanida could not exit the capsule because he was not able to disconnect his oxygen hose, which had frozen, from the gondola's oxygen supply.[6] He landed in a trash dump. During his third and final attempt, his face mask depressurized and the lack of oxygen severely damaged his brain, leaving him in a four-month coma before finally died. Ryan explains his motivation to tell this story by stating that "the story of Project Strato-Jump is very much a tale of its time" and as such is "one peopled with a cast of remarkable individuals, American originals all."[7] According to Ryan, 1966 "marks the end not just of a memorable man, but of a magnificent American dream, of an era," and it is on him, Ryan, to secure Piantanida's place in US American history.[8] Indeed, Ryan senses a weakening of Man's fatal attraction—not because of Piantanida's unsuccessful attempt to raise divinely and fall masterfully but, as the following paragraph shows, because not everyone felt as attracted to the American dream or the narrative that had always accompanied it:

> By the spring of 1966, not only fate, but the times, had caught up to Nick. On the morning of the day on which he set the unofficial manned-balloon altitude record, the New York Times carried on its front page two stories about seemingly unrelated events...In New York thirty-two demonstrators had halted traffic in Times Square during a sitdown protest

against the bombing of civilians in North Vietnam, and in Greenville, Mississippi, 110 civil rights workers and poor blacks had occupied a deactivated Air Force base and were forcibly evacuated . . . The changes—in America, in Western culture—would erase Strato-Jump from memory, as if a new collectivist radar was incapable of even discerning the solo act. Nick Piantanida had the misfortune to be cast as a pioneer just as the pioneering days, the cowboy days, of upper-altitude and space exploration were drawing to a close.[9]

Ryan's fears, however, are unsubstantiated. In fact, Piantanida and Baumgartner, or at least their names, not only have been set in literal stone but have secured a spot at the center of the world—and not just literally but even legally. In 1985, Jacques-André Istel, a Princeton graduate, parachutist, and retired investment banker, wrote a children's book—*Coe: the Good Dragon at the Center of the World*—in which he declared a city called "Felicity" the center of the world. A year later, he went on to found that city, named after his wife, on a 2,600-acre parcel of land he acquired in California's Sonoran Desert, nine miles west of Yuma, Arizona. He convinced both California's Imperial County and the French government (to be precise, its Institut national de l'information géographique et forestière) to designate a spot within the city as the Official Center of the World.[10] A great lover of monuments, Istel erected a pyramid right over the spot along with a chapel and a 2,100-foot triangle-shaped granite monument that together make up the Museum of History in Granite.[11] In addition to the histories of Arizona, California, and the United States of America, the monuments also contain The Hall of Fame of Parachuting. The two most recent engravings there are the names of Baumgartner and Piantanida, which is how they came to be present, or at least represented, at the center of the world. The existence of the Hall of Fame in the self-curated Museum of History is not too surprising; Istel, who worked closely with and financially supported Piantanida, is also known as the father of American skydiving (sport parachuting). In 1957, together with Lew Sanborn as vice president, Istel cofounded Parachutes Incorporated, the world's first parachute company, and introduced recreational free falling to the United States. Before this, the Civil Aeronautics Board had considered parachuting exclusively as a lifesaving emergency procedure for aviators in distress and strictly regulated all free falling.[12] After all, at least in the United States of America, the art of free falling was first and foremost a matter of war. However, Istel and Sanborn also taught the US Army's first free-fall course. Indeed, not only is Istel considered the man who brought more sophisticated free-falling techniques from France to the US generally but in 1957 he was also hired by the US Army to teach a group of selected military parachutists—a US special force—how to free fall in accordingly special, more advanced ways. Only two years later, as if no continuity exists, Istel published an article in the June issue of *Flying* magazine announcing that a "peaceful revolution . . . is taking place in the sky," as "aviation as a sport and youth movement is coming into its own." A "major motivating force for this revolution," he writes, is sport parachuting.[13] Echoing this sentiment, Ryan states that, thanks to Istel's endeavors, "free fall became quite literally a way of life."[14]

Another part of Istel's Museum of History in Granite is the History of Humanity, which consists of events, concepts, and ideas—curated by Istel— engraved on granite, running the gamut from the Big Bang to Barack Obama's presidency. The texts accompanying the panels are written by Istel in consultation with his wife. Panels five and eight, for example, are devoted to "our sun" and "our moon"—always making sure to use possessive adjectives; it is, after all, the history of humanity. Panel thirty is dedicated to "early concepts of law," featuring the Ten Commandments and excerpts of the Babylonian Code of Hammurabi from 1775 BC—law, too, is presented as a major human achievement. Panel 416, the "end panel," reveals the imagined future audience of Istel's educational project and furthermore provides the conceptual link that, as I will show, ties together the fatal attraction of Man with the will and power to free fall, the uncontrollability of physical forces, and the desire to reach the final frontier: "Unless we destroy ourselves, or succumb to a cosmic accident, our destiny should be set on a path to the stars." After a thick, white question mark, roughly the size of the quote, it reads in a slightly smaller font: "May distant descendants perhaps far from planet earth view our history with understanding and affection. Jacques-Andre Istel."[15] The questions that occupied Istel and his brothers-in-arms (and -in-free-fall) have not been answered, nor are they meant to be answered. They are always seeking another frontier, another occasion to ask: "How high could you go?" and "the complementary question: how far could you fall?"[16]

## Are Some Bodies Just Freer Than Others?

On October 14, 2012, Baumgartner, not yet honored with having his name engraved at the center of the world, was certainly the center of millions of viewers' attention. How high can he go and how far can he fall? This is as much a question of physics as it is of law, and even injustices. Baumgartner's record-breaking fall was not only, as the German comedian Jan Böhmermann satirically remarked, the realization of mankind's dream to "fall down from somewhere really high up."[17] Legally permitted to free fall to earth, that is, to fall from an atmosphere lacking air molecules and thus also air resistance, Baumgartner and the Red Bull team showed, once again, not only how human minds can make human bodies master physical forces but also how some bodies move more freely than others. Indeed, besides Baumgartner's navigation of legal, economic, political, and societal forces, as well as the challenge posed by gravitational force, drag force played a surprisingly major role in this event too. With a speed of up to 1,357.6 kilometers per hour (843.6 miles per hour) Baumgartner fell faster than expected. This led to a scientific investigation into why Baumgartner—an irregular body, with limbs, a space suit, and a backpack—fell faster than a smooth, even body would have. The difficulty lies in the calculation of fluid dynamics in the transonic range close to the sound barrier, where, as the physicist Ulrich Walter explains, various physical phenomena overlap: Air begins to act stiffly, forming shockwaves and causing turbulences that absorb energy and increase aerodynamic drag. However, under certain flow conditions, surface irregularities, such as those upon Baumgartner's body, actually reduce aerodynamic drag. In other words, it turns out that Baumgartner fell faster because apparently the sound barrier—a sonic wall—generates hardly any additional drag.[18] Baumgartner's fall was celebrated as yet another step by humanity toward previously unreachable frontiers. This success was further enhanced by the fact that Baumgartner's fall was even freer, that is, experienced even less resistance, than expected. *Are* some bodies, we might ask, *just* freer than others?

That different bodies are subject to different laws and move differently according to legal regimes is certainly not a new claim. Yet if bodies are understood in terms of Barad's agential realism, as discussed in the first part, then this claim requires law and legal thought to attend differently to how bodies matter and come to matter. Rather than bodies simply taking their place or being in the world, being situated or located in particular environments, Barad argues that bodies, "not merely 'human' bodies," are "integral 'parts' of, or dynamic reconfigurings of, what is."[19] Accordingly, a "robust theory of materialization of bodies" has to take account of "*how the body's materiality* (including, for example, its anatomy and physiology) *and other material forces as well* (including nonhuman ones) *actively matter to the processes of materialization.*"[20] Barad's claim must not be misunderstood as privileging the biological and physical over the social (or vice versa); it rather asserts that matter and meaning are intra-actively co-constituted, and all phenomena, including bodies and matter, themselves have their historicities. Furthermore, this argument does not allow for the assumption that forces, of whatever kind, simply act upon a preexisting body—let alone that bodies are products of external forces—as this would reiterate precisely the Newtonianism that Barad's work counters. Returning to the question of whether some bodies are just freer than others, it indicates rather that the constitution of bodies, along with their modes and possibilities of expressing concepts such as movement, freedom, mass, safety, breathability, attraction, acceleration, vulnerability, gravity, friction, resistance, and free fall, are inextricable from the force fields in which matter and meaning are co-constituted. Consequently, making-sense of law and legal force, even the force(s) of law, requires that legal thought become response-able to the differential materializations of bodies, understood not as effects of abstract legal orders, but matterphorically as entangled with matter(s) and force fields in their historicity.

In terms of the expression of movement or moveability (as the ability either to move or be moved) and how it relates to the matter(s) and force(s) of law, Baumgartner's fall reveals these entanglements even further. It goes without saying that for Baumgartner to even physically get to a point from where he could fall as freely as he did, he had to traverse multiple legal regimes and territories: Getting from Switzerland to New Mexico, for example, entails crossing various national and international regimes. Before and after the fall, he stands as an alien on US territory, and thereby within US sovereignty, subject to US law, state law, and local rules and restrictions. But even without his feet making contact with the ground, he moves horizontally across different airspaces and ultimately falls vertically through (legally regulated) spaces and regimes. Baumgartner holds EU citizenship, which facilitates easy crossing of national and international borders, and is supported by a multibillion dollar company; his fall was also registered by various tracking and recording technologies, which create and contribute to an image of a frontier-breaking hero. Moreover, his body was protected by a customized "personal life-support system," upholding his physical condition and biological functions in otherwise "hostile" space. It is no wonder then that his body seems to move almost *naturally* through all legal regimes.

Upon landing back on his feet and immediately falling to his knees, Baumgartner expresses his desire to "hug the whole world," evoking a sense of uniformity, undifferentiation, and wholeness; this utterly contradicts his strongly held political stances, which include his call for restricting refugee rights and enforcing impermeable border controls. Ironically, the man who fell the freest and felt the least resistance believes that not every*body* should move as freely. Baumgartner, as well as Red Bull CEO Materschitz, who passed away in 2022, openly supports far-right politicians—even those banned from entering the UK or classified as right-wing extremist entities in various European countries. In an open letter, Baumgartner also publicly warns against the alleged danger arising from "hundreds of thousands of refugees INFILTRATING—even if unarmed—our country" and reminds the state of its duty to "protect its own people and ensure the safety within its territory." He further states that "politics and correctness are as opposed as Islam and Christianity"—which is why, he claims, we (addressing mainly his Austrian, German, and Swiss audience) should not be too quick to give up our identity by "mixing it" with an "utterly different religion and ideology." Finally, Baumgartner recommends the Hungarian far-right prime minister Viktor Orbán for the Nobel Peace Prize as he is doing the "only right thing" by protecting "his land and his people" (sein Land und sein Volk) against refugees seeking asylum in Europe.[21] The latter should not surprise us; Baumgartner has repeatedly endorsed far-right politicians, including the 2016 Freedom Party of Austria (FPÖ) presidential candidate Norbert Hofer and the alt-right leader of the Identitarian Movement Austria Martin Sellner, who was recently in the news for potential links to the New Zealand Mosque attacks and for having put a swastika on a synagogue at the age of seventeen.[22] Sellner has been permanently banned from entering the UK because, according to a letter from the Home Office, he poses a "genuine, present and sufficiently serious threat to the UK's interests of preventing social harm, countering extremism and protecting shared values."[23] In July 2019, Germany's domestic intelligence agency (BfV) classified the Identitarian Movement as a right-extremist monitoring entity, stating that the movement "ultimately aims to exclude people of non-European origin from democratic participation and to discriminate against them in a way that infringes on their human dignity."[24]

When the Austrian private TV station *Servus TV* invited the Identitarian Movement leader Martin Sellner to participate in a 2016 discussion entitled "Radical Youth—How Dangerous Are Our Muslims?," three of the other invitees canceled their participation upon learning about Sellner's invitation. On the other hand, Baumgartner not only endorsed the station's decision to invite Sellner but also praised *Servus TV*, celebrating a "historical day for television." After the discussion was aired, he described Sellner as a "young, intelligent discussion partner, who convinces with politeness, eloquence and good arguments."[25] *Servus TV* is owned by none other than Red Bull, whose aforementioned CEO is also known for harsh remarks about refugees seeking asylum in countries of the European Union. In a 2017 interview published in the Austrian newspaper *Kleine Zeitung*, Mateschitz, back then already the richest Austrian, confirms his exuberant material wealth, which includes various real estate properties and even a South Sea island. He remarks that "it is true that I have a foible for beautiful and unique places in which I find pleasure, but I also look well after them." The implicit claim that his acquisition of "places"— that is, land property, which by nature excludes anyone else from its use—also serves the purpose of preservation takes on even stronger significance later in the interview when he complains about the "inability of mastering the surge of incoming refugees."[26] He boldly states that: "As with everything, there exists a *critical mass*. On a weekend, 20,000 hikers and mountain bikers in the Hohe Tauern National Park, that works. In that case, the chamoises squeeze together a little bit. But if the 20,000 become 200,000, or even two million, then the whole thing breaks. We have to understand that not only natural resources, but *all resources are finite: energy, water, food, air, medical care, everything, including Earth itself.*"[27]

In case there is any doubt about the evocation of tremendous force in his statement, it should be emphasized that "critical mass" is a term in nuclear physics denoting "the minimum mass or size of fissile material required in a nuclear reactor, bomb, etc., to sustain a chain reaction." According to historian of science Alex Wellerstein, it was first used in 1941 "specifically to talk about whether masses of fissile material could be made to explode on demand and not before."[28] Whether or not Mateschitz intentionally uses the term metaphorically, his pairing of an allegedly immigrant-induced threat to dwindling material and life-sustaining resources with destruction (kaputt gehen) by evoking the language of force and control is alarming. Materschitz's claim to take good care of *his* places and his warning that "Earth itself" is finite have to be understood in precisely the context in which they are uttered: as he says, the "flow of immigrants and migrations of peoples" (Auswanderungsströme und Völkerwanderungen). When asked what, in his opinion, governments could have done better in the past, he unflinchingly evokes legal and police force: "Of course borders should have been closed and strictly controlled, no question."[29]

The same year *Servus TV* invited Sellner, Baumgartner caused another uproar by sharing a phrase taken from a right-wing platform on his Facebook account, stating that "a country where you can't go fishing without permission, but where people can cross our borders with no passports, can only be run by IDIOTS."[30] Besides demanding unrestricted access to marine and land territory, he claims to feel "most at home in the air"—and thus he resents the outrage he caused in 2014 after violating the airspace of the Munich airport, where he entered without permission and flew his helicopter below 700 meters. And yet, legally, Baumgartner is currently less "at home in air" than in Switzerland, since this is where he changed his residence to from Austria—a shift he describes as an act of "tax optimization" rather than "tax flight."[31] Baumgartner certainly desires to move freely. It seems all too reasonable to him that for certain bodies, law *just is* what ensures unrestricted access to terranean, marine, and aerial spaces, and what makes certain modes of existence possible at the necessary expense of others.

It might not be a surprise, then, that the man whose achievement was to fall the fastest, to break through the sound barrier with a specifically engineered life-support system that sustains all biological functions, including respiration, is unwilling to accept what he might deem a limit to his power. Indeed, in May 2020, months after the WHO had declared COVID-19 a global pandemic, and a day on which it reported 4,789,205 COVID-19 infections globally with a death toll of 318,789 (3,318 on that day alone),[32] Baumgartner claimed that the wearing of face masks worsens the spread of the virus, demanded that the Austrian government abolish any obligation to wear masks, and ended his post with #keineMachtdenMasken (#nopowertomasks). Four days and 415,303 registered infections later,[33] Baumgartner, in yet another Facebook post, called for political disobedience (politischer Ungehorsam), ironically misusing a quote by Albert Einstein.[34] After receiving criticism for voicing his suspicion about COVID-19 vaccinations in July 2021, he defended his opinion by stating that "it is a free world and we all should respect that."[35]

By tracing the histories and forces of the supersonic fall in relation to microfascistic (physical and digital) expressions that aim to determine which bodies and modes of existence enjoy the possibility, if not the right, to move, fly, fall, and be, it becomes evident that breaths, as material-discursive and entangled phenomena, matter differently. These modes of mattering require close attention in order to address injustices deeply intertwined with how the world materializes. Furthermore, this investigation exposes a fact that will be further elaborated throughout this matterphorical case study: Practices and modes of governing—from legal and financial systems to oxygen supply, cultural production, technologies, and knowledge production—are significantly expressed, performed, constructed, and established by practices outside of, yet entangled with, the academic and public discourses of law and legal rights—notably by bodies falling to the ground *differently*.

# 8

# Home of the Mind

I AM REPRESENTING,

THEREFORE I AM

"Down with the law of gravity!" Down with all of nature's laws! —A., ARCH-ANARCHY

Governments of the Industrial World, you weary giants of flesh and steel, I come from Cyberspace, the new home of Mind. —JOHN PERRY BARLOW, "A Declaration of the Independence of Cyberspace"

New life
What if I need one?
New apartments on the Moon
Be the first to move there soon
With space to breathe
And lots of room
Maybe not
Maybe not
Maybe not
Maybe not.
—JAN BLOMQVIST, "Maybe Not"

Red Bull CEO Mateschitz felt betrayed. Baumgartner did not keep his promise to initiate his epic fall with the words "This is the World of Red Bull" but decided to announce his going "home." Casting the energy drink as that which "gives wings" suggests that it can enable the performance of otherwise impossible stunts or the avoidance of inconvenient situations that would be, without drinking a can of Red Bull, deemed hopeless. The declaration of planet earth as the world of Red Bull would have gone even further. Energy drinks that vitalize mind and body, water bottled during the full moon (LunAqua), Red Bull mobile services, major sport sponsorships, Formula One racing teams (Red Bull Racing, Scuderia Alpha Tauri) and sport team ownerships, a media company (Red Bull Media), a television channel (*Servus TV*), and a wide array of other products and services—Red Bull has been proliferate in broadcasting its brand, including its logo, on earth. At some point, however, neither earth nor logos are enough, not even on spacesuits or capsules in the stratosphere. Not only are representations *of* outer space, such as science-fiction movies and images of planetary surfaces, in high demand, representation(s) *in* outer space, including advanced commercial space media, promise unprecedented profits. Seen from this angle, Baumgartner's refusal to mention Red Bull is indeed a missed opportunity for the company to prove that neither physical laws nor any other laws can stop its all-consuming hubris. In this context, too, Baumgartner's position at an altitude of 38,9694 kilometers (24,2145 miles) is crucial—not only because he falls from a zone of legal indeterminacy but also because he leaves the capsule right where commercial advertising eagerly strives to finally reach outer space.

For decades, companies have been trying to push legislation that would enable "space advertising," for which no clear definition currently exists in international space law. In addition, the absence of a clear spatial definition of where air space ends and outer space begins significantly informs discussion about, and regulation of, commercial activities. From the early 1990s onwards, proposals to use outer space for commercial advertising purposes proliferated—from "space billboards," arranged formations of bright satellites, and "moonvertising" to the placement of a Pizza Hut logo on the surface of a Proton rocket; from the use of the space stations MIR and the International Space Station (ISS) to record advertisements by the Israeli milk company Tnuva and Japanese instant-noodle company Nissin's "Space Ram."[1] Questions about the legitimacy of space advertising have been addressed by international committees and legal theorists, who have concluded that, currently, the "making of commercial representation for the purpose of promoting sales of goods and services in outer space" is not explicitly prohibited by international space law.[2] While all space activities must comply with space law—such as the Outer Space Treaty (OST) and the Moon Treaty—space advertising is subject to supervision by individual states and affected by national space-law regimes.[3] However, increasing pressure from the private sector, looking to reach an unprecedented number of potential consumers from space, has led to the implementation of a distinction between nonobtrusive and obtrusive advertising. While the former includes, for example, "product placement used in broadcasts from space objects, logos on uniforms of persons in space, and logos on space objects themselves" and is, as such, not prohibited, classification becomes more intricate with the latter. Defined as any "advertising in outer space that is capable of being recognized by a human being on the surface of the Earth without the aid of a telescope or other technological device,"[4] the concept of obtrusive advertising presents various challenges for space law as well as international and national law: light pollution, space debris, obstruction of astronomical observations, and the right of consumers *not* to read advertisements.[5] What is more, according to the OST and the Moon Treaty, all activities in outer space have to "benefit all mankind," which is, as one proponent of space advertising admits, difficult to claim when it comes to commercial representation.[6] The push to allow space advertising is markedly provoked by the evolving space tourism industry and other profit-driven endeavors aimed at finally breaking into space; it is also, as will become clearer, often tied to particular ideas of self-governance. With this in mind, the argument that space advertisement—for example, a Coca-Cola logo shining brightly from the night sky—benefits all mankind seems even more ironic. Outer space, celebrated as the "next frontier" for commercial activity, including advertising, is indeed a space hospitable to all kinds of tropes, including irony.[7]

For example, a few years ago, the Russian start-up StartRocket founded by Vladilen Sitnikov planned on placing space billboards made of CubSats in low earth orbit. The project, which was scheduled to be test launched in 2021, was called the "Orbital Display." It was to consist of an array of satellites—each with a thirty-foot reflective sail that would function as a pixel and reflect the light of the sun—orbiting the planet and displaying logos visible from any part of the earth for a certain amount of time. StarRocket's website prompted its visitors to "Say Hi to New Media" and claimed to "create a media" with "potential audience coverage of 7 billion people on the planet." The project's aesthetics, based on light pollution, space debris, the potential danger of collision with other space objects, and commerce, were justified with the start-up's assessment that "space has to be beautiful." And what could be more beautiful, we read, than "the best brands," with which "our sky will amaze us every night?" The prospect of space debris did not resonate well. The company experienced enough pushback to shift their focus to another frontier by working with foam as "a new way of space exploration." To "make this dream come true," the website states—and here lies the irony—space needs to be cleaned up first. This is why StarRocket's most immediate goal is now the development and launch of the "Foam Debris Catcher," described as "a spacecraft series created to help us clear the way to distant frontiers for future generations within a few years." Once the path is cleared, foam will, per the claim, become a "new building material for outer space colonies."[8]

It is, however, not just irony that permeates outer space—humor seems to travel lightly too. On April 1, 2021, Musk, known for his abundant use of memes on X (then called Twitter), tweeted that "SpaceX is going to put a literal Dogecoin to the literal Moon," a post that mobilized the Dogecoin (DOGE) hype at the time. The "doge" in Dogecoin refers to a famous internet meme featuring a Shiba Inu dog and colorful text in Comic Sans font, usually in English and deliberately unrestrained by English grammar. The meme's eventual becoming as a decentralized, blockchain-based cryptocurrency, founded by software engineers Billy Markus and Jackson Palmer, was originally meant as a joke: "a parody of all the 'serious' clone coins that were trying so hard to differentiate themselves" yet seemed exactly the same and, as Markus claims, were "all touting how they were going to become worth zillions and take over the galaxy."[9] DOGE's reputation as the people's coin that, despite being a highly volatile meme currency, is believed to be going to the moon speaks to the affective functioning of memes.[10]

Sensing and sensemaking operate accordingly. Musk does not specify what he *means* by "literal Dogecoin." Cryptocurrencies, as the name suggests, are digital currencies. Physical (doge)coins do not have currency value, are not considered a currency, and function only as memorabilia.[11] Yet despite the suggestive date of its publication, Musk's post must not be mistaken for an April Fools' joke. As early as February 2021, Musk tweeted an image that showed, among other things, the Doge staking a Dogecoin flag into the surface of the Moon. The image sat between the words "literally" (above the image) and "on the actual Moon" (below the image). In crypto slang, "to the moon" expresses an investor's expectation that a cryptocoin (or an asset) will increase significantly in value, that is, rise up until it reaches a peak ("mooning.")

Only a few months later, the planning for a mission to send DOGE to the moon was underway. On May 9, the Canadian intellectual property, manufacturing, and logistics firm Geometric Energy Corporation (GEC) announced

Elon Musk ✔ @elonmusk · Feb 24, 2021
Literally

Elon Musk ✔
@elonmusk

On the actual moon

2:10 PM · Feb 24, 2021

♡ 180.9K    ♡ 5.9K    ⬆ Share this Tweet

FIGURE 8.1. Twitter (now X) posting by Elon Musk on
February 24, 2021.

the DOGE-1 mission to the moon, a collaboration with SpaceX to launch a
forty-kilogram CubeSat as a rideshare on a Falcon 9 lunar payload mission
initially scheduled for June 2022. The fact that the mission was fully paid for
in DOGE, as GEC's Chief Executive Officer Samuel Reid claims, "solidified
DOGE as a unit of account for lunar business in the space sector." SpaceX
itself also expressed its excitement "to launch DOGE-1 to the Moon!"
The mission itself will, as SpaceX Vice President of Commercial Sales Tom
Ochinero states, "demonstrate the application of cryptocurrency beyond
Earth orbit and set the foundation for interplanetary commerce." In a rather
cryptic manner, the press release also states that "additional payload space will
be allocated to include digital art in the form of space plaques provided by
GeometricLabs Corporation and Geometric Gaming Corporation."[12] Soon,
more information was revealed about GEC's plans, which have less to do with
digital art than with venturing into space advertisement. Indeed, the aim is
to have space display screens (1200 × 1080 OLED displays) in not one but,
according to GEC's website, twelve large U6 CubeSats, which orbit the earth
and the moon.[13] These space billboards will display custom commercial advertisements, "visible globally via in-space cameras" and "streamed to sponsored

websites, YouTube, and other social media channels." Because these displays will not be visible from earth, the advertisement is categorized as nonobtrusive, yet it still fosters the commercialization of outer space. The underlying business model builds on Web 3.0 (Web3),[14] more precisely, the XI Protocol: a distributed ledger technology (blockchain) solution that uses space infrastructure as its physical platform and "intends to establish an interplanetary platform" for providing financial incentives and participate in new economic models.[15] An important element includes the so-called "Space Tokens"—KAPPA, BETA, GAMMA, RHO, and XI—which are utility tokens that allow holders to rent pixels on a display in space.[16] BETA and RHO are used to claim, respectively, the x-coordinate and y-coordinate on the satellite screen, while GAMMA is used to claim the brightness and KAPPA the hue of the pixels on the screen. The XI token is used to claim time (duration) for the advertisement. In other words, the tokens can be purchased and then converted to rights to control pixels on the satellite.[17]

Pushing the next technological and commercial frontier by using Web3 technologies to enable space-enhanced advertisement, the mission is cast within the extropist and transhumanist narrative of humanity becoming a multiplanetary species. As the Geometric Space Corporation states on its website, "the intermediary steps include optimizing current technology and expanding the commercial space market," but the company further claims their "ultimate ambition is to contribute to the creation of a multiplanetary civilization."[18]

Other than StartRocket, which decided to introduce a cleaner image (so to speak) only upon receiving pushback, the DOGE-1 mission builds, at least when it comes to public perception and hype, on Dogecoin as a highly affective meme and its specific connotations. In an example that demonstrates how, returning to the inaugural issue of the *Extropy* journal, "mindfucking can be used to open up people's brains to extropian perspectives,"[19] Musk followed up on the GEC press release with an X post stating "SpaceX launching satellite Doge-1 to the moon next year—Mission paid for in Doge—1st crypto in space—1st meme in space. To the moooooonnn!!" (@elonmusk, May 9, 2021, 6:41 p.m. EDT). The post also contains a YouTube link to the popular "Dogecoin Song," which starts with:

> What is the only crypto currency that you should invest? Dogecoin!
> It's the people's currency, you know it's the best. Please mine!
> The most fun and ironic crypto that is for sure. So crypto!
> We're taking Dogecoin to the moon! To the moon![20]

DOGE's value went up because it's fun(ny) and soon fell rather seriously, disappointing and (actually) devaluing the monetarized hopes that fueled (parts of) the people's coin's upward trajectory. The game, of course, is not over. Hodl the void, even if everything around you is falling.

If you can afford the steepest falls, you will rise—to the moon, and beyond. As David Gokhshtein, an American influencer and crypto investor, optimistically writes in a reply to Musk's post, "DOGE to the moon! This is a lifestyle" (@davidgokhshtein, May 9, 2021, 6:47 p.m.). That this lifestyle—albeit ironic and humorous, even a joke, a finance comedy sold as a liberation story for and by the people—is inextricable from modes of existing and even governing is expressed by yet another phrase that permeates online fora, appears in X posts, and features prominently on memes: "Long live the doge!"

### Frontierism, Ashtronauts, Frozen Assets

Unsurprisingly, Red Bull, in the business of seeking out new frontiers to tackle, has already entered the crypto-world with NFTs and Futurocoin.[21] And as we will see, while not every*body* needs to stay alive to increase the company's market value, its unique marketing strategy prefers bodies that not only are alive but that even outperform their physical capacities. As the description of the Red Bull podcast "How to Be Superhuman?" reveals, the goal is to "travel to the very limits of human endurance" and to "do things most mortals couldn't even dream of"; in short, to become "superhuman."[22] The company's representation—all but representational—capitalizes on audacious individuals jockeying to wear the company's logo while defying physical limits in *breath-taking* stunts—falling from a space capsule, slacklining over a canyon, moving between gliders in midair, or jumping on a motorcycle to the top of the Vegas Arc de Triomphe and then back down again. The Red Bull Air Force's declared mission, for example, is "to push the limits of human flight" and to "continually rethink what is possible and envision a world where all people live with boldness."[23] The ever new high-quality action-camera equipment capable of producing the desired (commercial) representation and the imminent chance of death the athletes are willing to take in the name of the brand, its philosophy, and ultimately its profit are both crucial elements of this strategy. As one German journalist surmised in an interview with *Die Zeit*, the company's calculations always included the possibility of Baumgartner's death; it would have, per the journalist, led to a tremendous increase in Red Bull's market value due to increased publicity.[24] As of now, ten bold Red Bull athletes have lost their lives in their attempts to perform their dangerous stunts, among them the

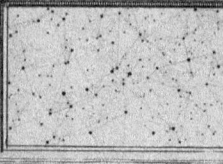

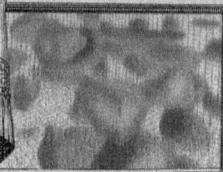

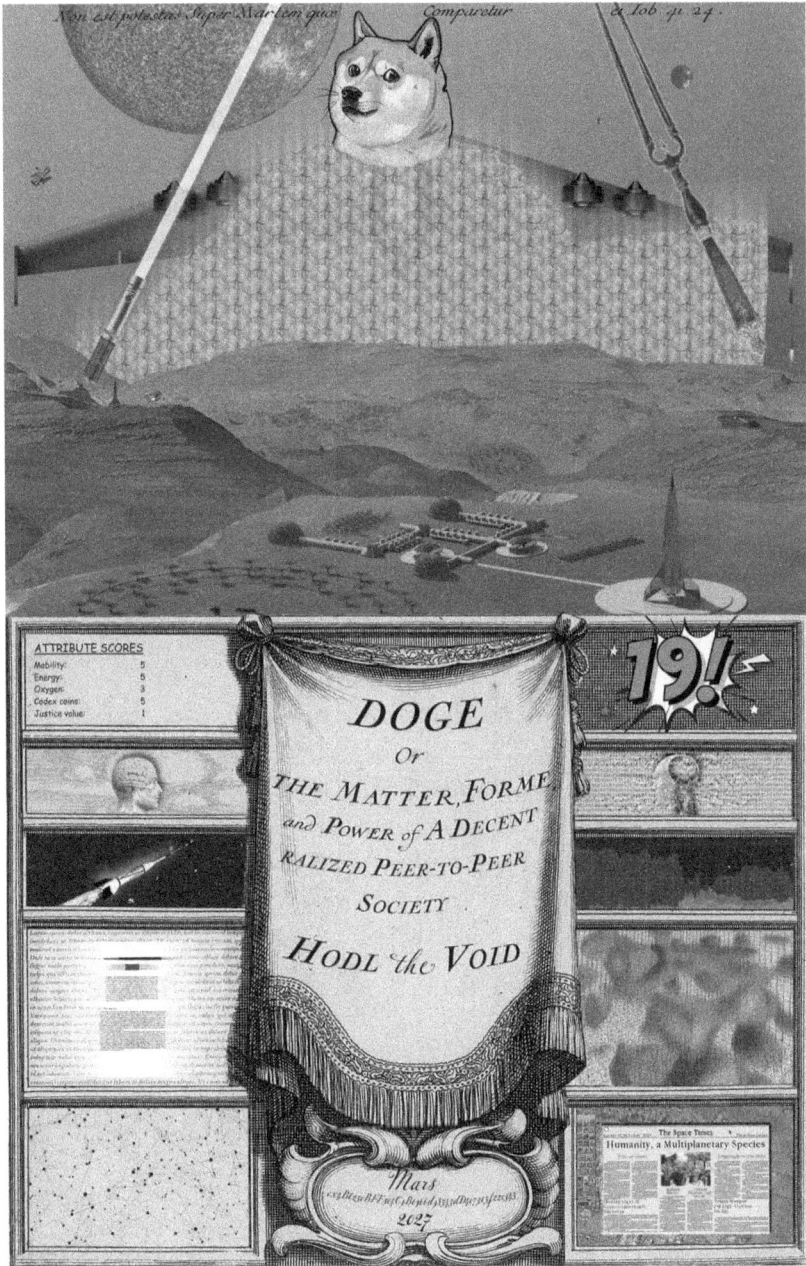

FIGURES 8.2 AND 8.3. Two NFTs from the Hodlthevoid NFT collection, an experiment with governance NFTs by LoPh+.

Russian building, antenna, span, and earth (BASE) jumper Valery Rozov, who crashed while jumping off Mount Ama Dablam in Nepal; the US-American freestyle snowmobile rider Caleb Moore, who suffered a fatal crash from performing a backflip during the Winter X Games; Vince Reffet, a skydiver and BASE jumper who died during a "jetman" training session in Dubai and was, according to Red Bull's obituary, "at home in the air"; and Shane McConkey, the Canadian skier and BASE jumper who lost his life in the Italian Dolomites in the course of a documentary production by Red Bull Media House.[25] The trailer for *McConkey* heroically speaks of breaking limits, living free, and taking risks to change the world. The documentary, replete with Red Bull logo placements, was released years after the athlete's death with the ironic subtitle, "You have one life. Live it." The fiction renders bodies unbound by any restrictions and posits a world in which no border, wall, demarcation, or other physical limit can stop those adventurous and daring enough. Breaking and transgressing limits is, in this imagination, enabled by an increased energy level—160 grams of caffeine and 54 grams of sugar per can promises to deliver what is needed to transcend limits: "Red Bull gives you wings."

Baumgartner's stratosphere jump exemplified the slogan, reinscribing both the narrative of the human pioneer, daring enough to prove critics wrong, and that of the visionary entrepreneur making the impossible possible. Both narratives significantly drive space literature, travel, commerce, and colonization. Taken together, the matterphorical case study of the jump reveals, once again, how colonialism and frontier logic determine notions of progress. Take, for example, Musk's claim that humanity will either cease to exist or become a digital, multiplanetary species, "ensuring the light of consciousness is not extinguished."[26] As mentioned before, the SpaceX founder and CEO is willing to take financial risks but never physical ones. He acknowledges the dangers and high probability of death of the first journeys to Mars and admits that he will likely not be on board, as he wants to see his kids grow up. However, Musk is certain that people will sign up for the mission because, "just as with the establishment of English colonies, there are people who love that"; they "want to be pioneers."[27]

This language is not coincidental. In his essay on the half-life of empire in outer space, Peter Redfield explains that in the aftermath of the twentieth century, space exploration advocates "constitute perhaps the last unabashed enthusiasts of imperialism," employing the term "colony" and "cheerfully describing conquest, settlement and expansion," for "theirs is a Columbus of exploration, nation building and risk taking, not of invasion, domination and genocide."[28] As Fraser MacDonald states, the colonization of space is by no means a decisive or transcendent break from the past but "merely an extension of long-standing regimes of power."[29] Marina Benjamin, examining the history of the United States' conception of outer space, shows that it was "always a metaphorical extension of the American West," imagining, as she writes, "a handful of brave pioneers charging routes through unknown and perhaps dangerous new territory, followed by colonists who begin homesteading on the edge of the known world."[30] In one of the most influential books on American space colonization, *The High Frontier*, Gerard K. O'Neill calls the prospect of space settlement a chance for "rare, talented individuals to create their own small worlds of home and family, as was so easy a century ago in our America as it expanded into a new frontier."[31] And he speculates "with some supporting evidence" that these space settlers will most likely be from the United States, Canada, Australia, and other former colonies, as descendants of former colonies have a "greater desire for travel and change." It is clear, he writes, that "the early settlers in space will be exciting people: restless, inquiring, independent; quite possibly more hard-driving and possessed by more 'creative discontent' than their kin in the Old World."[32]

Inherently expansionist colonial imaginaries and their related legal concepts, such as *terra nullius* and homesteading, are, alongside anarchocapitalist endeavors to "keep law short" or replace legal orders by architectures of code,[33] penetrating atmospheres and space(s), thereby determining what modes of existence the future can hold. And while it is primarily nations and empires that are associated with colonies, corporations, too, were in charge of all aspects of life in a colony, including, as Persson writes, possessing "the right to decide and enforce local laws and to judge in local matters."[34] With the emergence of NewSpace, private corporations aim to undermine legal systems, attacking laws that restrict resource extraction and the unlimited gain in profit for the wealthy few.[35] At the same time, NewSpace proponents push for significant legal modifications to secure and widen their own property rights. Indeed, the greatest challenge that space law—especially the OST and the Moon Treaty—currently presents for private corporations and investors is that it prohibits national and private appropriation, as well as sovereignty claims, thereby discouraging high investments;[36] there are fewer possibilities, as proponents of the private space industry argue, to gain profit from outer-space activities. For example, the legal scholar David A. Collins calls for an "internationally recognized legal regime for property rights on Mars" because a lack thereof will "endanger financial investment both in reaching and then colonizing the planet."[37] Assuming that Mars is devoid of "value," he argues that the planet will be "rendered valuable by adding the characteristics of accessibility and habitability," which is why "the party who reaches and develops it first should be able to claim owner-ship."[38] The legal parallel he draws from the early settlement of the American frontier can be "readily analogized to a planet because both consist of unde-veloped, uninhabited physical space."[39] In the footnote to this analogy, he unflinchingly states that he is "ignoring, for the purposes of *comparison*, the aboriginal presence in the American west."[40] The injustices sealed within these concepts reemerge later in the essay when Collins acknowledges that in case of an "emergency" or "catastrophe" on earth, an international earth government might need to "seize land on Mars for public use" but would have to "compen-sate the original owner in the process." This "emergency trespass," he insists, "would not preclude the payment of reasonable compensation for use or damage to existing infrastructure."[41]

In addition to the nonappropriation principle, space law (like its terres-trial analogs, the UN Convention on the Law of the Seas [UNCLOS] and the Antarctica Treaty System [ATS]), is guided by a *Common Heritage doctrine*.[42] In other words, because outer space (and also the high seas) is considered *res communis*—property of all—it is not subject to private ownership.[43] Benefits from the extraction of resources must be shared with all.[44] However, recent in-struments in the laws of both the sea and outer space have been directed toward the assertion of a free-market principle, which makes it rather unlikely for de-veloping countries to expect much from the utilization of either space-based or deep-sea natural resources.[45] In addition, the United States, and subsequently also Luxembourg, has passed laws that give companies property rights to the space resources they extract. Although this has been met with resistance from space experts—as the OST states that "outer space and all non-man-made ob-jects it entails are subject to *international* regulation"—it demonstrates the ac-celerating efforts to commercialize and privatize outer space and its resources.[46]

In keeping with extropist desires, these developments materialize Jackson Turner's prognosis in his 1893 "The Significance of the Frontier in American History," according to which "American energy will continually demand a wider field for its exercise."[47] Such materializations are inseparable from what normativity is becoming in emerging legal concepts and modes of governance. Frontiers are force fields in which power seeks to transcend its limits; law, as a boundary-drawing practice, is present even in (or precisely through) its ab-sence. In fact, frontiers perform a questioning of law and normativity because they function as "in-between" spaces, or fields of indeterminacy, in which entrenched determinations are unsettled and the call for new ones rises.[48] Fron-tiers, as Anna Tsing notes, evolve "in the shifting terrain between legality and illegality, public and private ownership, brutal rape and passionate charisma, ethnic collaboration and hostility, violence and law, restoration and extermination," precisely because a frontier is "a zone of not yet—not yet mapped, 'not yet' regulated."[49] Or, as Mattias Borg Rasmussen and Christian Lund remark, "frontier dynamics dissolve existing social orders," such as "property systems, political jurisdictions, rights, and social contracts," and entail dynamisms of reordering and determination.[50] As history shows, the expansionist logic driven by "the discovery and invention of new resources" results in frontiers becoming simultaneously sites of horrendous destruction and potential abundance, of deterritorialization and reterritorialization, of indeterminacy and the desire to determine and order. Law's most significant eureka moments are accompanied by the rise and fall of power(s)—and with them, values, histories, bodies, borders, and particles.

It is instructive to read Turner's claim in relation to the aforementioned extropist desires. Let me thus return to the fifth issue (1990) of the *Extropy* journal, to a contribution published under the pseudonym "A." and titled "Arch-Anarchy." The essay starts with the following paragraph under the subheading "Call to Arms":

> Down with the law of gravity! By what right does it encounter my will? I have not pledged my allegiance to the law of gravity; I have learned to live under its force as one learns to live under a tyrant. Whatever gravity benefits, I want the freedom to deny its iron hand. Yet gravity reigns despite my complaints. "*No gravitation without representation!*" I shout. "Down with the law of gravity!" Down with *all* of nature's laws! Gravity, the electromagnetic force, the strong and weak nuclear forces—together they conspire to destroy human intelligence. Their evil leader? Entropy. Throw out the Four Forces! Down with Entropy! Down with *every* limitation! I call for the highest of all freedoms. Come, let us cast off *all* chains! We will make our own heaven. We will become our own gods. I call for perfect self-rule; I call for arch-anarchy![51]

In this imagination, arch-anarchy, which follows "individualist anarchism," is the "highest form of anarchy" as it not only rejects state laws but denies "the validity of every law—human or otherwise"—that gets in the way.[52] Natural laws are just "observed constants" that indicate not how the universe *must* behave but only "how it has been observed to behave by certain scientists in certain labs at certain times under certain conditions."[53] The author's conclusion is that "we have broken natural laws before and we can break them again"; doing so is simply a question of will and "mere technical details."[54] Those willing to join arch-anarchy will, aided by technology, together "cast down the laws of

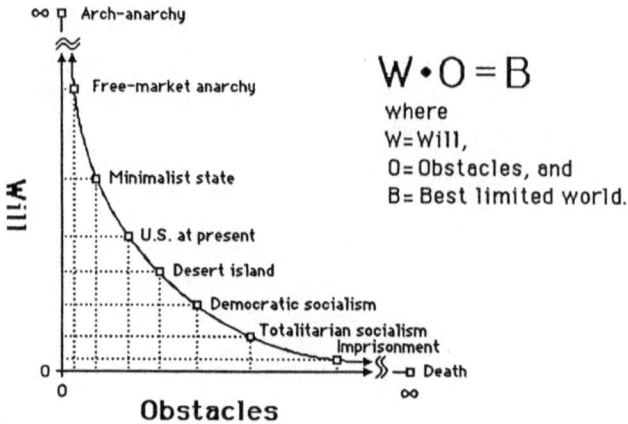

$$W \cdot O = B$$

where
W = Will,
O = Obstacles, and
B = Best limited world.

Graph labels (top to bottom): Arch-anarchy, Free-market anarchy, Minimalist state, U.S. at present, Desert island, Democratic socialism, Totalitarian socialism, Imprisonment, Death. Y-axis: Will. X-axis: Obstacles.

FIGURE 8.4. "Graph of reality" indicating where arch-anarchy falls in relation to systems of governance. *Extropy*, issue 5, Winter 1990.

statists, moralists, nature, and logic" and thereby be given "the power to fly in defiance of gravity, live forever," create their "own universes," and become God—which would still, the author notes, not be enough. The "ultimate goal" is to become "the singular, perfect, omnipotent power of God."[55]

Arch-anarchy, as the rule of an individual will unobstructed by "obstacles," is contrasted with death as the absence of will and the presence of stasis. Put differently, the desired state of lawlessness (or, depending on the point of view, of being personally unconstrained by law) is placed in opposition to a state of undesired absolute regulation. The formula, read together with the author's claim to have divided the universe "into two opposing forces: my will and obstacles to my will," where ("I") refers to "a particular pattern of information, a set of data and processing rules,"[56] mobilizes a Newtonian notion of force as the ideal (self-)governing principle. Those unattracted to arch-anarchy are considered obstacles and even enemies to the arch-anarchist will, and consequences are adumbrated: "If godhood doesn't appeal to you, then you aren't much like me anyhow and it probably isn't in my self-interest to drag you into heaven."[57] Although Reilly Jones, another *Extropy* contributor and transhumanist, uses a different tone in his essay "The History of Extropian Thought," published six years later, both the onto-epistemological assumptions and the extropist desires mobilizing them remain central. Extropy, he writes, will finally create the freedom to "become so exceptional as to be a law in ourselves," to "free ourselves 'from the earth, rise into the infinite, leave the bonds of the body, and circle in the universe of space amongst the stars,'" to "make 'every solid in the universe . . . become fluid' on the approach of our mind," to "become 'a perpetual living mirror of the universe,'" and finally, even most importantly, to "'be everything' and 'be it forever.'"[58]

## Before the Law: Legal Falling, Legal Standing

After falling 38,969.4 meters (127,852 feet) from the sky, Baumgartner lands near Roswell, New Mexico, which lies at an altitude of 1,089 meters (3,573 feet). The team celebrates. Without a doubt, besides the lack of any legal restrictions on falling freely, the advanced design and creation of a wearable system that upheld Baumgartner's life functions, regulated temperature, and kept air pressure in check made it possible both to break the sound barrier and to keep his record-breaking fall on record—digitally, analogically, and even engraved in stone. The spacesuit's four layers consisted of "'breathable' Gore-Tex, and heat- and flame-resistant Nomex" that retain air pressure, prevent tissues from turning into gas, and protect the body from heat and temperature extremes.[59] In order to ensure that Baumgartner would at no point or moment be out of breath, his composite pressure helmet was equipped with an oxygen regulator that provided him with 100% oxygen "to breathe from various sources"; specifically, "a liquefied oxygen source on the ground before launch, from the capsule's liquefied system when he's onboard, and from a pair of high-pressure gaseous oxygen cylinders during the freefall descent."[60] Needless to say, even with a personal life-support system, falling from the stratosphere is rather unpleasant. In an interview in 2013 with the *Financial Times* (conducted shortly after he received an award at the 2013 Laureus World Sports Awards and featuring a photo of him posing on the only 396-meter-high Sugarloaf Mountain in Rio de Janeiro, Brazil), Baumgartner recalls what it meant for him to breathe not only pure but too pure air in an environment with not only less (legal and physical) pressure but none at all. It is, he reveals, "such a hostile environment up there," as "everything's dangerous when you're in space because there's no pressure." The life-saving suit is uncomfortable. "You are locked in your own little world," breathing "100 per cent oxygen for an hour before you start," and, when ascending and falling, "you can hear nothing but yourself breathing for hours and hours." The "horrible sound of me breathing," he continues, was recorded just once by the team and played to people who tore off their headphones after only five minutes. To emphasize the challenge of breathing under these conditions, he adds that "you have to use a lot of force to breathe, it's not natural"; it is "like breathing through a pillow . . . you keep thinking, I'm not getting enough air, so you panic, freak out and want to tear the thing off." He declares that he never wants to be in one of those suits again and that if he ever has kids who decide to emulate his fall he would be "horrified."[61] This might also explain one of his earlier statements, in which he recalled his first few seconds back on earthly ground. "When I finally landed, and got out of the suit," he states, "it felt like the prison doors opened now and I can just walk away."[62] Ironically (or not), this statement is reminiscent of a Red Bull TV advertisement in which a prisoner, shackled by ball and chain, his outfit fashionably striped, drinks a can of Red Bull, approaches two guards, kisses one goodbye and, once Red Bull gives him wings, simply flies off into freedom—into the air.[63]

Whether falling or rising, Baumgartner is not only representing Red Bull but also represent*able*: He can seek legal representation and, in some cases, even represent himself before national and international courts. Successful in breaking the sound barrier and pushing his physical and psychological limits, it remains to be asked whether Baumgartner rose high enough to transcend even his legal status—that is, the position he occupies in relation to law and within the legal systems and regimes he falls through. If so, what role does altitude, and with it the decrease in oxygen and the shifts in breathability, temperature, and resistance, play in delimiting concepts and their reach? Do concepts differentiate between air—a mixture of gases—that suffocates and air on which wings beat?[64] And if so, what are the material-discursive cuts that separate one from the other? To reiterate, despite the media declaring Baumgartner's fall a "space jump," a "plunge from outer space," a "jump from the edge of space," and a "leap to Earth," the indeterminacy of when airspace ends and outer space begins (or vice versa) leaves it ambiguous whether he actually fell from space.[65] As Konstantin Kakaes puts it succinctly in the *MIT Technology Review* Space Issue, "what's space in one context is still atmosphere in another, depending on whether you're a satellite, an astronaut, or a would-be space tourist."[66] The indeterminacy unsettles not only the legal concept of *sovereignty* (and *ownership*) but also *legal status*: who or what Baumgartner *is* in terms of law and, I would add, its entanglements. Clearly, Baumgartner is not a satellite, but should he be considered an astronaut? Does this question matter?

In space law, the potential consequences of the legal status of astronauts are significant. "All States Parties" to the OST, notwithstanding conflicts, "shall render to them all possible assistance in the event of accident, distress, or emergency landing on the territory of another State Party or on the high seas."[67] Similarly, the 1968 Agreement on the Rescue of Astronauts, the Return of Astronauts, and the Return of Objects Launched into Outer Space (Rescue Agreement) lays out clear rules about rescuing "personnel of a spacecraft" in case of accident, emergency, distress, or unintended landing. If the personnel of a spacecraft land, under the mentioned circumstances, "in territory under the jurisdiction of a Contracting Party, it shall immediately take all possible steps to rescue them and render them all necessary assistance."[68] What is more, the OST considers astronauts "envoys of mankind," representing not only themselves or their countries but earth's population as a whole.[69] This is also the case because astronauts—on a mission to reach scientific and technological frontiers while discovering new spaces, often uninhabitable for human bodies— are conceptually descendants of a particular legal subject; that is, a particular holder of rights and obligations. Indeed, critical and feminist legal theory has exposed the Cartesianism fundamental to the construction of the legal subject, which, although a crucial concept for the development of rights, is modeled after Man: the disembodied, property-owning subject.[70] Facilitated by Kant's and Hegel's legal theories, which propagate a "decontextualization of the subject from the world of objects, including the body," the legal subject developed as disembodied, "more like a cypher" than a natural person.[71]

Yet the body exists—a fact that legal thought cannot ignore. Legal thought is informed by Newtonian notions of force and Cartesian dualism. On one hand, it imagines law as acting on bodies, presupposing embodiment as the foundation of its enforcement. On the other hand, when law addresses bodies, it relies on a series of constructions (for example, "the body as machine, property, autonomous, bounded, commodified").[72] Bodies thus "never completely disappear in law" but hover between disembodiment and materiality, displaying their contradictory existence and occupying a rather complex position while still considered as playing "no essential role in the provision of meaning."[73] What is more, the conception of the physical world as an accumulation of objects and the concept of the body as having its own self-enclosed, bounded nature are historically inextricable from the concept of ownership and property.[74] In short, these ontological assumptions, splitting minds from bodies and subjects from objects, are prerequisites for upholding the concept of "rationalistic juridical subjectivity fundamental to the operationalization of a quintessentially contractual legal and political order."[75] Because legal sense is not only considered detached from the body but also positioned "outside" of the physical world, law and legal concepts are assumed not be co-constituted by them, which reinforces their position "beyond to the world, in the ethereal, immaterial realm of word, of mind, of reason."[76] Transcending matter and becoming representative, representable, and representing is not only a requirement for falling, swimming, thinking, and breathing safely but also part of the dynamism of the very concepts (such as social contract, rational subject, individual, sovereignty) on which liberal political and legal orders—entangled with colonial and imperial histories—rely. As Redfield states, "once a few white men moved beyond the *atmosphere*, they became newly, artificially human by virtue of the nonhuman space around them, cast as universal representatives by virtue of their transcendent, hazardous location."[77] Pointing to terrestrial histories of colonization, Laura Zhang importantly notes that "both astronauts and future space settlers are considered explorers and heralds from Earth."[78] The legal status and matters of bodies—that is, the possibilities of their physical and social existence—are intimately related. Moving beyond breathable atmospheres can go either way. Indeed, while some bodies transcend to represent, or to legitimately find an image justifying their existence, others are left to drown without their non/existence being represented, let alone accounted for. Indeed, refugees and asylum seekers trying to cross the Mediterranean Sea without sufficient technological, nutritional, or legal support, for example, are not considered worthy of representing humanity as envoys but are rather pictured as a "wave" or "surge" infiltrating, flooding, and destroying Europe.[79] The fact that some bodies breath liquid oxygen while others are left to breathe water, or to combat breathe,[80] cannot be seen as material-discursively disentangled from legal status and the question of representability, let alone representationalism as a mode of thought and sensemaking. What is unrepresented and unrepresentable—or what refuses representation—breathes differently, if it breathes at all.

Legally, it can be argued that Baumgartner does not qualify as an astronaut or as part of the personnel of a spacecraft. And yet, even though he only fell from twenty-four miles up—about twenty-six miles lower than the NASA and US Military and National Advisory Committee on Aeronautics (NACA) altitude for astronaut classification—the particularities of the fall expose the limits and indeterminacies of legal concepts, including those of legal forms of existence. As private companies, such as SpaceX, founded by Musk, Blue Origin, founded by Amazon CEO Jeff Bezos, or Virgin Galactic, founded by Richard Branson, venture into space travel, concepts such as "commercial astronaut," "space tourist," and "space colonialist," which cannot be defined in reference to either the OST or the Rescue Agreement, are increasingly being discussed and shaped.[81] And if, as astronomer Chris Impey claims, the first issue beyond survival for humans in space will be their legal status, then it is crucial to attend to the ways in which physically leaving the earth, and perhaps even digitally traveling between atmospheres, enables different material-discursive practices of legal sensemaking and concepting.[82] Can we still take it for granted that altered atmospheric conditions do not change or affect what legal concepts can *do*? What happens to the concept of rights, including the human right to life, when oxygen is low and hydrogen is scarce? What legal concepts will evolve from oxygen-deprived environments, be they Mars or metaverses, and what does this mean for modes of existence on earth, where the ozone layer in the upper atmosphere is thinning? Put more generally, how will legal concepts express *mattering* before (perhaps even after) the law?

To demonstrate why such considerations are essential, a glimpse into the literature on the imagined early days of outer-space colonization—when resources will be scarce, and oxygen and water are expected to become the most precious currencies—proves helpful.[83] Addressing the fact that terraforming and colonizing environments that are uninhabitable for human bodies will present enormous hardship, the physicist Adam H. Stevens suggests "penal colonization," which, according to him, "raises an interesting prospect for future extra-terrestrial colonization." While it may seem unsavory, he continues, "cheap labour from criminal elements of society might be worth considering for the early phases of extra-terrestrial settlements, at least while living conditions and security of essential resources remained poor during the early stages of colonization."[84] With regard to the political and legal challenges, astrobiologist Charles Cockell warns that, in outer space, people will eventually control the processes between oxygen atoms and the people that breathe them. Oxygen, he continues, "provides the most thorough example of something that creates opportunities for the concentration of power and wealth."[85] Highly technological and expensive systems, scholars argue, are needed to secure the basic survival of human and nonhuman lives, first of all by providing—and restricting, distributing, trading, and selling—breathable air.[86] Fearing a legal and economic environment in which corporations hold a monopoly on oxygen, Cockell calls for "a type of oxygen independence"; he argues that because of the harsh, extraterrestrial physical conditions, combined with a strong free-market ideology, "outer space has a tendency to encourage totalitarianism."[87] That these calls should be taken seriously is clear in the arguments brought forward by those who embrace free-market liberalism in outer space. Stevens, for example, anticipates oxygen scarcity in early colonies and bases his speculations on the availability of oxygen on each respective planet. In a lunar colony, oxygen will be drastically scarce, and "whatever entity controls production of oxygen will have major economic and political power," whereas on Mars, where oxygen can be extracted from the carbon dioxide atmosphere, "more complex market forces" will control the oxygen supply.[88] One mentioned possibility would be a so-called "low-oxygen diet" for nonwealthy settler colonialists who contribute less to the space economy and yet consume oxygen. Colonists who "cannot afford to purchase their required air and water for the month" could opt to "purchase less their needed oxygen and ration themselves to breathe lower than the standard 21% oxygen mix."[89] Stevens admits that this might "create a complex new social class based on nothing more than how much you breathe." Despite the framing of the essay as an investigation into the harsh conditions of the onset of space colonization, Stevens concludes confidently that the growth of space colonies and any systems involved must be carefully managed, at least until these colonies develop to a stage where oxygen is abundant.[90]

Finally, to reiterate that the aim here is not to declare enemies, pinpoint villains, or oversimplify complicities, let me point to the fact that there is no consensus even within alliances. While the image of thought—its representationalism and expansionism, its ontological dualism, and the consequent superiority of the mind—remains fundamental, there have been major disagreements among the extropians, and also with and among cypherpunks, on the role of law in their movements (especially regarding the US Constitution and forms of private law).[91] Consequently, consensus is lacking also when it comes to the question of Man falling and rising, of vital forces and fatal attractions, of bodies arresting the self that eagerly desires to be freed, and of corpses burned and frozen on their final rise. Laws apply, jurisdictions draw lines, and contracts are signed—and canceled. Timothy Leary, an extropist thinker, fervid proponent of consciousness expansion, and cult figure of the psychedelic and human-potential movement in the United States, was a longtime supporter of the cryonic project but decided to eventually cancel his contracts with the Alcor Life Extension Foundation and CryoCare; to use his friend Barlow's words, he "died unashamed and having, as usual, a great time."[92] While Leary was posthumously criticized in the seventeenth *Extropy* issue for choosing "heat death" (incineration) over cryonic preservation and was thus accused of having died from "inattention to reason," he had in fact chosen to rise high, to "ride the light into space."[93] Indeed, in what became the first commercial burial in outer space (sponsored by Celestis and launched from the Canary Islands), his remains, together with the ashes of twenty-four other men, including Gerald O'Neill, a physics professor at Princeton University in the 1970s and major proponent of space colonization, were carried to low earth orbit. Seven grams of ashes were encased in individual lipstick-sized aluminum capsules, which were expected to make at least 8,600 orbits before being recremated upon reentering earth's atmosphere.[94] The cost of the flight was $4,800 "per *asht*-ronaut."[95] In contrast, yet entangled with Leary in many ways, Hal Finney, a cypherpunk, renowned cryptographer, and contributor to the *Extropy* journal, made different arrangements. Finney, a blockchain pioneer who received the first transaction on the Bitcoin blockchain from Satoshi Nakamoto (a pseudonym used by the author(s) of the Bitcoin white paper and creator(s) of Bitcoin), decided to direct his last transaction to Alcor and become, in exchange for Bitcoins and life insurance payments, Alcor's 128th "patient." After having been "pronounced legally dead" on August 28, 2014, at 8:50 a.m., his corpse was, per his instructions, prepared for cryonic storage. His blood and bodily floods were drained and replaced with a mix of chemicals that Alcor refers to as M-22, a composition of dimethyl sulfoxide (22.305 percent), formamide (12.858 percent), ethylene glycol (16.837 percent), N-methylformamide (3 percent), 3-methoxy-1,2-propanediol (4 percent), polyvinyl pyrrolidone K12 (2.8 percent), and X-1000 and Z-1000 ice blockers (3 percent) that prevents crystallization of cell membranes. Cooled down to minus 196 degrees Celsius, the corpse was then, according to Alcor's cryoconservation protocol, stored in an aluminum pod inside a tank filled with 450 liters of liquid nitrogen "until the day when repair and revival may be possible."[96] Currently, the Alcor Whole-Body Preservation costs between $200,000 and $220,000.[97] This, too, is a lifestyle: Long live the mind!

What this matterphorical case study already shows is that beneath seemingly solid legal concepts lies not only uncertainty but indeterminacy. Indeterminacy, whether understood in an epistemological or ontological sense, has historically accompanied frontier spaces; it has been met there less with a commitment to attend to it than with a desire to determine, to unroot existing structures and set the rules for new ones. As Benjamin Ross explains, transhumanist thought addresses limits and ontological uncertainty with a firm conviction in epistemological certainty, rooted in its belief in the unlimited potential of human enhancement through technology and (artificial) intelligence.[98] Politically driven by anarchocapitalist and extropist goals, this approach produces normative structures based on private property, market competition, and stakeholder value, free from government and international law restrictions. In international law, indeterminacy has been a major structural concern in assessing enforceability. This indeterminacy is often perceived as "a result of conflictual and antagonistic relationality,"[99] closely following the Schmittan logic of existential negation and friend-enemy dynamics. However, postcolonial scholarship has pointed out that international law is, in fact, founded upon conflict. This manifests not only literally, in terms of the Thirty Years' War and the subsequent emergence of the Westphalian model of sovereignty, but also by means of what Antony Anghie calls a "dynamic of difference," which establishes legitimacy only through the negation of the illegitimate—that is, those in need of the supervision of rational judgment and judgmental reasoning.

Put differently, separability and oppositional binarism, mirrored in the friend/ enemy, civilized/uncivilized, and sovereign/nonsovereign dichotomies, are essential to the sense and meaning-making mechanisms that underlie international law and its concepts. Either way, although differently so, sensemaking and normativity-constructing mechanisms that underlie colonial responses to indeterminacy are guided by separability and oppositional binarism and mirrored and reflected through the friend/enemy, civilized/uncivilized, supply/ demand, and sovereign/nonsovereign dichotomies.

The potential of a matterphorical approach lies in attending to indeterminacy affirmatively and creatively: giving rather than taking (a) place and creating conditions for alternative concepts that enable different modes of existence. Matterphorics involves first tracing legal concepts to their points of indeterminacy—where assumptions about their presumed immateriality collapse and the power of their matterphorical constitution is revealed—then offering a synaesethics of legal concepting. This approach recognizes the power of what legal thought and concepts can do and responsibly delineates their potential and limitations, especially in the most enticing and dangerous spaces: frontiers.

Baumgartner's fall was sensationally declared a "plunge from outer space" and a "jump from the edge of space." These announcements suggest that Man can now fall not only from the sky and the heavens but also from outer space. In terms of law, however, it is not clear whether he actually fell from space. As of now, there exists no clear definition in international law of either outer space or airspace. Consequently, there is also no clear definition of where the boundary between these two inherently different legal concepts, and corresponding different regimes and forces, lies. As Francis Lyall and Paul B. Larsen point out, various theories and suggestions on how to determine where "space as a matter of law" begins have already been put forward.[100] Most approaches to the establishing of this boundary focus directly on physical elements and forces: in relation to the earth's gravity (so called "gravispheres"), the various layers of atmosphere, or certain altitudes at which, for example, atmospheric drag becomes noticeable or "where aerodynamic lift yields to centrifugal force."[101] Other suggestions pertain to the different effects of spheres and spaces on human bodies (e.g., lack of oxygen, evaporation of fluids, ultraviolet radiation, cosmic rays) or nonhuman bodies (e.g., airplanes and rockets).[102] But each approach comes with significant consequences for all kinds of bodies and modes of existence. Linking the upper limit of airspace to the subjacent states' ability to exercise control within their airspace would, as some experts suggest, inevitably leads to boundary drawings that vary across the globe; this risks revealing "the inequality of Nation-States' abilities to project force."[103] Altitude-based approaches, too, negotiate boundaries in accordance with forces and their interplay (or intraplay) with bodies (whether human, airplanes, rockets, etc.). Examples include the boundary used by NASA Mission Control, which is set at an altitude of 76 miles (122 kilometers) and is understood as the point of reentry "at which atmospheric drag becomes noticeable."[104] Another example is the von Kármán line, used by the Fédération Aéronautique Internationale (FAI), the International Air Sports Federation, and the US National Aeronautic Association, which lies at an altitude of 62.5 miles (100 kilometers) and designates "the point where aerodynamic lift yields to centrifugal force."[105] As an article published in *National Geographic* summarizes, outer space is often considered to begin "at the point where orbital dynamic forces become more important than aerodynamic forces."[106] The indeterminacy of the two concepts (airspace and outer space), entangled with force(s) and matter(s), affects and co-constitutes other legal concepts in turn. Indeed, what is also blurred in this vertical legal zone of indetermination is one of the major concepts in international law: state sovereignty, or the right of a state to exercise its jurisdiction. Airspace is sovereign space, which means that "a state is entitled to enforce its sovereignty by shooting down an unauthorised intruder as an exercise of sovereignty," a practice that has occurred repeatedly.[107] In contrast, outer space is governed by the corpus of space law, which explicitly prohibits sovereignty and property claims. "Outer space, including the Moon and other celestial bodies," Article 2 of the OST states, "is not subject to national appropriation by claim of sovereignty, by means of use or occupation, or by any other means." Similarly, the Moon Treaty declares that the moon and "other celestial bodies within the solar system, other than the Earth," are "not subject to national appropriation by any claim of sovereignty, by means of use or occupation, or by any other means."[108] The reluctance of major spacefaring states to resolve this indeterminacy speaks to the complicity as well as to the entanglements of legal concepts (as *matterphorical expressions*) with political, scientific, historical, technological, and other desires to expand states' airspace as far as possible into outer space—and to avoid establishing a boundary to the upper limit of their force.

The boundary between airspace and outer space is not an anomaly but rather makes visible how legal thought seeks to protect its own solidity. Attending to this indeterminate boundary already demonstrates what will be further discussed later in this part of the book: how matter and physical forces— gravitational force, nuclear force, drag force, dynamic and aerodynamic forces—pose challenges to the constitution and expression of legal concepts (such as force, sovereignty, outer space, airspace, the high seas) rooted in unquestioned onto-epistemological assumptions. Such assumptions are powerful. Legal concepts favoring solid ground and land mass facilitate the establishment of fixed boundaries, borders, and property relations. Legal space and time become divisible, measurable, and delineable. Newtonian and Cartesian understandings of space, time, and scale, considered "a natural measure of space and time in terms of intervals or distances,"[109] as well as the Cartesian dualism that separates subject from object, are powerful tools—more precisely, tools of power. Proprietary lines can be drawn in the soil and on maps. Representations hold and inscribe power. Schmitt, whose work is currently experiencing a splendid comeback across disciplines, feared the decline of European legal power as law moved beyond being construed as solely terranean and land-bound and suddenly appeared in relation to oceans and airspace—when the various states of matter and their entanglement with the power and force relations underlying legal concepts could no longer be ignored.[110] For Schmitt, law itself has its legitimate roots in land appropriation and colonization. An understanding of law attentive to other states of matter—and other matters of states, let alone other states, let alone alternatives to State thought—is, in his view, a threat to international law and therefore to central European sovereignty (and supremacy).[111] Resistance to states of matter, to matter(s) of law and force, is a defense mechanism that mobilizes enormous amounts of power, force, and energy.[112] It is no coincidence that, historically, the very materiality of the high seas, the deep seabed, Antarctica, and outer space—all at some point imagined as new "frontiers"—has presented challenges to law and legal order. It is rather difficult to draw lines in the sky, determine borders in the air, or define boundaries when the state of matter is not solid but fluid or, in this case, gas.

Continuing this matterphorical case study, the final chapter in this section returns to look for what happens to legal thought and concepts, rooted in notions of territory and land governance, when there is neither solid ground nor land in sight. Where do other states of matter fall in this legal theory? Can Man fall for power from lower altitudes, descending even deeper below the sea level, or is there a natural limit to his fatal attraction? In what follows, the entanglement of falls will lead from space to the deepest point in earth's oceans, from liquid oxygen to held breaths, from canons to cannons, from ballads to ballistics, from literature to law (and vice versa), from vertigos to vortices, from kings and territories to waves and forcefields, from naked ideas in a tub to Baumgartner's pressure helmet, and back again to the frontier. Humanity, so the story (and also the poem) goes, is under high pressure as consciousness desires to expand—downwards, upwards, outwards.

# 9

# Under Pressure

## FALLING DEEPER FOR POWER

What if he who gives you air gives you air so rarefied, or compressed, or pure, or polluted, or . . . or . . . that he, in effect, gives you death? —LUCE IRIGARAY, *The Forgetting of Air in Martin Heidegger*

I can't breathe.
I can't breathe.
I can't breathe.
I can't breathe.
I can't breathe.
I can't breathe.
I can't breathe.
I can't breathe.
I can't breathe.
I can't breathe.
I can't breathe.
—ASHELY YATES (@BROWNBLAZE), quoting Eric Garner, as Daniel Pantaleo choked him to death.

One hundred ninety years before Red Bull launched its first energy drink in Austria and two hundred fifteen years before Baumgartner fell from the sky, another man is said to be standing at an edge; he is looking bravely but frightened into depths that will require him to master another set of forces with brazenness and physical strength. Although this venture is exactly the kind that Red Bull would gladly sponsor, wings are not what this daredevil needs. Even falls are entangled; this man, too, decided to fall in front of a live audience that watches with curious and emphatic fear for his life. But in his case, this is just half of the story, for his stunt consists of not only jumping from a cliff into unruly water but also diving into the dark depths of the ocean to retrieve a golden chalice—proving that he is capable of negotiating and mastering various forces. This is, of course, also long before the Austrian diver Herbert Nitsch failed in breaking his own free-diving record (214 meters/702 feet) in 2012 after suffering from nitrogen narcosis and severe decompression sickness and even longer before Red Bull Media House released the documentary *Herbert Nitsch: Back from the Abyss.*

And yet there is already a lot of pressure. First of all, everyone is looking, including the king and sovereign himself, who posed the challenge that the young man hesitantly accepted. If the sovereign poses a challenge, one legitimately feels pressure to perform as requested. In fact, for some bodies, there are not many options besides complying or disappearing—dead or alive—into the legal spaces the sovereign's force cannot reach—in this case, the depths of the ocean. Before the invention of aviation, let alone of space flight, the Roman private law maxim "cuius est solum, eius est usque ad coelum"—which would also come to guide international law until the OST entered into force—stated that "whoever owns land, it is theirs up to the heavens and down to hell."[1] This maxim, however, pertained to land mass, not the ocean. Stemming especially from warfare and maritime trade, the question of how far the power of the sovereign reaches from their territory into the sea had, at the time of this man's imminent fall, already been an ongoing legal debate. Here, too, high pressure is involved. Among the suggestions as to where to set this boundary were a sixty-mile mark, a one-hundred-mile mark, a two-day sailing distance, and even the range of human vision.[2] Later, in 1702, the Dutch jurist Cornelius van Bynkershoek famously rejected these contemplations and argued in favor of tying the limit of sovereignty to the power to control the territory through the force of arms: "*The power of the land properly ends where the force of arms ends. Therefore the sea can be considered subject as far as the range of cannon extends.*"[3] What half a century later became known as the "cannon-shot rule" provides yet another example of the intra-actions of power, force, and matter in the constitution of law and legal meaning. Cannons—up to the mid-nineteenth century—used black powder, a chemical explosive consisting of a mixture of potassium nitrate ($KNO_3$, 75 percent), wood charcoal (C, 15 percent), and sulfur (S, 10 percent). The high pressure of a compressed gas, resulting from combustion in a confined place, propels the cannonball, which—depending on angle, acceleration, velocity, and air resistance, among other factors—travels a certain distance. Potassium nitrate is crucial because, when ignition energy is applied, it functions as the oxidizing agent, releasing oxygen that combines with the carbon of the charcoal to produce heat and gases.[4] The explosion heat of black powder is 250 joules per gram, the specific gas volume is 280 milliliters per gram, and the explosion temperature is approximately 2,400 degrees Celsius.[5] Various forces are apparent in the reaction. The ignition results from friction; in the barrel, force is produced by the expanding gases and by atmospheric pressure; outside the barrel, the cannonball is subject to gravity and viscous drag. Besides high pressure and force, the power to affect a body (in this context, primarily a ship) plays a major role in legal considerations too. At the end of the eighteenth century, territory was understood to be acquired by occupation rather than naturally belonging to a piece of land or sea, and thus it was inextricably linked to "the exercise of sufficient power to reduce the terrain to possession and to exclude others from possessing it."[6] As Natalie Klein writes, the law of the sea in general, and the concept of a terrestrial sea in particular, developed in relation to maritime and military interests, "particularly in favor of the states with the greatest power in this regard."[7] The entanglement of physical force(s), military technology, and chemistry with naval and sovereign powers and their relationship to physical states of matter (solid land, fluid oceans) was constitutive of the legal rule that has been influential for the subsequent three-mile rule and, ultimately, the five main zones in the contemporary law of the sea (the United Nations Convention on the Law of the Sea).[8]

Even if, as was the case at that point, the king's force is imagined as reaching out horizontally along the curved trajectory of a cannonball, the pressures of the depths of the sea, where sovereign power is difficult to enforce, are not to be underestimated. In fact, the young man's life depends to a significant degree on how his body encounters pressure. There is atmospheric pressure, the force exerted on a surface by the air above as gravity pulls it to earth, and also hydrostatic pressure, the force per unit area exerted by liquid on an object, also due to the force of gravity. There are the tides, mainly depending on the moon's gravitational force or, as Owein Jones puts it, "expressions of the interplay of many profound forces, particularly the rotation and tilt of the earth and the relational movements of the heavenly bodies."[9] There are hydraulic forces, maelstroms, waves breaking and diffracting—and also, everyone is watching.

If that weren't enough, the diver—as the German writer and poet Friedrich Schiller called the brazen young man in the eponymous ballad—is under moral pressure too. Perhaps this is where Schiller empathizes with him. "The Diver" ("Der Taucher") was the first ballad written during the famous "Year of the Ballads" of 1797—a time of literary theorization and creative competition between Schiller and the German writer Johann Wolfgang von Goethe. It was written only a few years after the French Revolution, amidst attempts to allegedly rescue German literature from its alleged aesthetic decline and, additionally, under the scrutiny of Goethe, who pressured Schiller to let the diver die. In fact, while Schiller was still working on the poem, Goethe ended a letter to him with "farewell"—literally "live just and well" (Leben Sie recht und wohl)—and added, "let the diver drown as soon as possible." Goethe's reasoning is telling, as it exposes which forces he conjures up to display his power to affect the bodies he imagines: "It's not bad, because I send my couples into the fire and rescue them from the fire. It makes sense that your hero chose the opposite element."[10] Here, Goethe is referring to "The God and the Bayadere" ("Der Gott und die Bajadere"), a ballad he was writing around that time. Yet in the very same year—1797—he also wrote "The Sorcerer's Apprentice" ("Der Zauberlehrling"), one of his most famous ballads, in which an apprentice overestimates his power by conjuring up magical force and risks not only drowning but flooding his master's house: "Brood of hell, you're not a mortal! / Shall the entire house go under? / Over threshold over portal / Streams of water rush and thunder."[11] Time is pressing, forces are out of control, enchanted brooms keep on fetching water, thresholds are washed away by currents and the rising water level. Fears here pertain less to diving than to drowning and flooding. The apprentice eventually admits his mistake; he should have obeyed because only the sorcerer can, by means of language, command a force that powerful: "Off they run, till wet and wetter / Hall and steps immersed are lying. / What a flood that naught can fetter! / Lord and master, hear me crying!— / Ah, he comes excited. Sir, my need is sore. / Spirits that I've cited / My commands ignore."[12]

Questions of obedience and compliance are tricky. Who can ignore commands and who has to—or is made to—comply? How do power and force(s) relate? Decisions have to be made, and they matter. "*To obey! To rule!—Immense, vertiginous gap ... To obey! To rule!—To be, or not to be!*," Schiller writes elsewhere.[13] The diver hesitates. On the one hand, there is the king on whose territory the diver stands, standing for the king's law, his wealth and property (including the golden chalice), and his power to affect monetary, material, and societal relations. On the other, there is the ocean, historically depicted both as an unruly, uncontrollable, and indivisible surface and as a dark and unknown depth—a legal space outside the king's territory and thus, as the poem implies, under the governance of the immortals or gods. Between them lies the free fall of the diver who, falling from solid land territory into a differently structured legal space, assesses which pressures his body can withstand and which will eventually break it. He decides to fall, diving after the golden chalice, but does so with the hope of returning to land even more respected and welcome than before his test of courage and loyalty to the crown.

However, there are also different, incomparable kinds of forces and laws that, when it comes to falling and diving for the crown and after the chalice, are nonetheless entangled with those of the sovereign. For one, the physical challenges the diver must face are significant. The deeper he goes, the more the pressure increases; Boyle's law states that if the temperature remains constant, the volume of a gas is inversely proportional to the absolute pressure—the volume of the air spaces in the body decreases.[14] According to Henry's law, the partial pressure of the air inside the lungs increases, and a greater concentration of gases, including oxygen, appears in the diver's lungs. Gas molecules diffuse, that is, they go from areas of higher concentration to lower concentration until an equilibrium is reached. This means that the deeper the diver goes, the more oxygen will be present in the blood. When ascending, the volume changes again, and oxygen comes back to the lungs. According to Archimedes's principle, the decrease in the volume of air spaces also effects the body's buoyancy. As we are told in the ninth book of Vitruvius's *De architectura*, Archimedes discovered the principle during an investigation into whether or not a contractor (a goldsmith) had broken the law and committed a theft against the sovereign by diluting the gold he received to manufacture the king's crown with silver. "While the case was still on his mind," Archimedes "happened to go to the bath, and on getting into a tub observed that the more his body sank into it the more water ran out over the tub." Because, we read, this taught him how to "explain the case in question," he "jumped out of the tub and rushed home naked, crying with a loud voice that he had found what he was seeking."[15] In any case, Archimedes's principle states that "any body wholly or partially submerged in a fluid is buoyed up by a force equal to the weight of the fluid displaced."[16] With the decrease in air volume in the body, the body becomes less buoyant. Buoyancy force also depends on density, and because seawater has a greater density than freshwater, it is easier to float and harder to descend into its depths.[17] However, some bodies *naturally* ascend and descend faster in dense and fluid mediums than others; the diver's first attempt turns out well. After narrowly evading being drawn into the depths of maelstroms and torrents, the diver eventually surfaces: "Then breathes he deeply, then breathes he long."[18]

## $H_2O\ L_4W$

That the diver in Schiller's ballad surfaces and can breathe again is, the text states, testimony less to his physical condition or mastery of physical forces or pressures than to the fact that he, by a sign of God to whom he prayed, found a grip on a craggy reef. From there he could also reach the chalice, hanging on sharp corals that prohibited its "fall into bottomlessness."[19] As we read, it was an upward force that, luckily, carried his body back up. Schiller himself never saw the sea. In a letter, he confesses to Goethe that he studied the motions and falling of water by watching a watermill—which, as a side note, uses hydropower to generate energy and facilitate the production of various material goods—and kept his descriptions close to Homer's descriptions of Charybdis: "It delights me quite a bit that, according to your observation, my description of the vortex is consistent with the phenomenon," Schiller writes.[20] The diver's narration of the unruly and dangerous sea, of monsters, creatures, maelstroms, and bottomless darkness, rests on representations, literary narrations, and spatial imaginations. Although described by Schiller with analogies and overly metaphorical language, the importance of breathing—in the diver's case, of the movement of air between the atmosphere and the lungs as the intake of oxygen and expelling of carbon dioxide—remains central. Indeed, the regained ability to breathe is related to the return not only to an environment where atmospheric pressure allows the volume of the diver's lungs to increase again but also to the sovereign and to the territorial space, the land and ground as opposed to "bottomlessness" and the uncontrollable play of oceanic forces, associated with his rule: "Long live the king! Let all those be glad / Who breathe in the light of the sky! / For below all is fearful, of moment sad; / Let not man to tempt the immortals e'er try."[21]

If the relationship between the sovereign's rule and the ability of certain bodies to breathe, as exposed here, is to be considered poetic, it is because poetics is anything but representational. In fact, as Glissant writes, under the conditions set by the thought of an empire, poetic thought seeks "the really liveable world" beneath the fantasy of domination. Poetics is creative and productive rather than descriptive or representative.[22] Poetics, Glissant explains, "aims for the space of differences—not exclusion, but, where difference is realized in going beyond."[23] And if difference is considered an onto-epistemological expression, as I have argued, then poetics is indeed "all the material dances, errant wandering/wonderings, diffractings, entanglings, reconfigurings happening in the field of desiring."[24] What men are poets, Richard Feynman asks in his lectures on physics, "who can speak of Jupiter if he were like a man, but if he is an immense spinning sphere of methane and ammonia must be silent?"[25] And, we might further inquire, what thinkers are legal theorists who can speak of constitutions but render oxygen, among other matters, negligible? As we have already seen, the question of how law relates to oxygen, to which forms and what kinds of oxygen—as a commodity in outer space, in its molecular structure as oxidizer that accepts electrons and fosters combustion in high-tech spaceships as well as cannons, as that which fuels wildfires, as what upholds all life functions from blood circulation to breathing and energy production—inhabits, even if unasked and unaddressed, legal thought and theory. In approaching the question of whether we should continue to accept laws, and even an understanding of law, that claim to exist even conceptually without oxygen, Luce Irigaray's *The Forgetting of Air in Martin Heidegger* remains instructive. Here Irigaray gives a *matterphorical* account of the indispensable relation between air—therefore also oxygen, for air is a mixture of gases including nitrogen ($N_2$), oxygen ($O_2$), argon ($Ar$), neon ($Ne$), helium ($He$), methane ($CH_4$), krypton ($Kr$), hydrogen ($H_2$), nitrous oxide ($N_2O$), and xenon ($Xe$)[26]—and philosophical thought. Irigaray shows how Western metaphysics has established a particular relationship to physics and matter that is oriented toward solids and fundamentally relies on the forgetting of air; that is, of fluids and gases. This mode of thought "always supposes, in some manner, a solid crust from which to raise a construction" and thus also "a physics that gives privilege to, or at least that would have constituted, the solid plane."[27] Although Irigaray mainly refers to Heidegger's work here, the specific mode of thought characterized by forgetfulness of matter and reliance upon a solid crust for conceptual construction is prevalent in legal thought too.[28] Importantly, Irigaray argues that it is exactly the forgetting of the materiality of air that constitutes the power of dominant thought: "On air, he nourishes himself; in air, he is housed; thanks to air, he can move about, can exercise a faculty for action, can manifest himself, can see and speak." And yet, the "aerial matter remains unthought by the philosopher."[29] The philosopher's Cartesianism makes him believe he can transcend the matter from which he evolved and which he inhabits. There is, Irigaray writes, a "vacuum" or "chasm at the origin of their thought's appropriation," which is created by such thinkers "using up the air for telling without ever telling of air itself."[30] In aiming to challenge the Cartesian word/world separation and the notion of logos that is, Irigaray demonstrates, inextricably tied to it, she astutely asks if air, a "fluid truth," is *thinkable*.[31] It is a matterphorical question. As such, it questions how a mode of thought—underpinned by the assumption that being is deprived of oxygen and situated in a vacuum (assumed to be empty)—could ever create the possibility to exist, live, and breathe for anyone, especially for those who don't matter (human and non-human) within this system. In relation to law, where matters of life, nonlife, and death are stipulated, we are prompted to ask once again whether an ethically tenable legal theory can afford to be forgetful, if not negligent, of matter. This includes the materiality of air, its specific physical properties, and its molecular, atomic, and in-elementary constitution. Irigaray, for her part, cautions against the destructiveness of subjecting matter to the alleged power of the mind. Ultimately, the forgetting of air's materiality and the appropriation of air, actually and conceptually, by means of disembodied thought cannot withstand the pressure exerted by the matter(s) at stake. Indeed, the power (*pouvoir*) of this materiality does not vanish. If philosophy forgets, then it will be, she warns, the "techno-physics" (*techno-physique*) that takes it upon itself to remind us: "for instance, through the splitting of the atom."[32]

How, we might ask, does the power and force of splitting atoms remind us of the air's materiality? Irigaray does not specify further here, but we do know that upon the detonation of a nuclear fission bomb, more than 50 percent of the energy is released as a blast wave formed by the initial, rapid expansion of a fireball that severely compresses the surrounding atmosphere. The airburst height of the atomic bombs dropped on Hiroshima and Nagasaki, about five hundred to six hundred meters above the ground, was deliberately chosen to maximize the destructive blast. Due to the enormously high temperatures and pressures at the point of detonation, the hot gaseous residues move outward at very high velocities radially from the center of the explosion. The resulting shock wave, or blast wave, initially travels at 30 km/second—one hundred times the speed of sound in normal air—compressing the air and heating it up to thirty thousand degrees Celsius.[33] Most of the material damage from a nuclear airburst is caused by a combination of static overpressure, which is the sharp increase in pressure due to atmospheric compression, and dynamic pressure, which results from the drag forces exerted by the blast winds. These blast winds form because the blast wave transfers its energy to the molecules of the surrounding air, setting them in motion. All objects, regardless of their structure or size, in the path of the blast wave will be subjected to the dynamic pressure loading or drag forces of the blast winds. Spacetimemattering, to speak with Barad, is "shattered, torn, broken, into dis/connected pieces . . . vaporized, dispersed, made particulate, whisked away on the breeze," when, in an instance of inconceivable endurance, "the whole world is downwind."[34]

The diver, not yet fallen from grace, relates differently to air. Continuously pushed to his limits, he can only hold his breath for so long. Understandably, he is more than grateful for the ability to surface again—to breathe legally and freely under the crown, with solid ground under his feet. The sovereign, however, is not satisfied—there is more to know, deeper to dive: "If thou'lt try once again, and bring word to me / What thou saw'st in the nethermost depths of sea."[35] The diver falls for power. Both the pressure of the proof of obedience requested by the king and the attractive reward he offers—marriage with the king's daughter and therefore the possibility to fall legally into power—impel him to repeat his stunt. Baumgartner, on the other hand, declares that he is not a "repeated offender" (*Wiederholungstäter*) in an interview conducted two years after his fall from the stratosphere. Rather than falling, Baumgartner explains, his "new task" is "giving lectures for managers all over the world."[36] Of course, comparison fails. Baumgartner fell for profit, the diver for political power: different forces, different powers, different matters. What remains, however, is the question of boundaries, limits, and, as we will see, frontiers.

The boundary between land and sea, through which the diver falls not only once but twice, is one that, as the literary scholar Gerhard Neumann argues, is omnipresent in Schiller's work, especially in his fragments. It creates, he further states, a "political topology" that emerges from the boundary between land and sea, between solid law and fluid lawlessness: territorial orders based on solid ground and the sea as a space of extraterritorial existence.[37] The notion of the sea as uncontrollable and unruly space, which informs the Western understanding of law, has a long history rooted in European legal thought and the development of international law. At the same time, however, the ocean functioned for some bodies, those freer than others, as a "space of connection and an arena of mobility."[38] The history of the international law of the sea—regulations pertaining to, among other things, marine movement and transportation, property rights, territory, resource extraction—is predominantly based in European powers and legal traditions. Even decisions to treat the oceans as commons are, as Becky Mansfield argues, consistent with attempts to foster mercantilism, exploration, colonial expansion, and cold-war military maneuvering. Representation of oceans as an immaterial, abstract space of movement that defies the drawing of lines, settlement, and territorialization constructed them as "open for the smooth movement of capital, resources and militaries" yet also as the outside that confirms and defines the territorial, solid inside from which law and sovereign power evolve.[39] Aware of the legal, political, and ideological implications of the land-sea boundary, recent attempts in critical and critical-legal geography have challenged this dichotomy by emphasizing the materiality of the sea, countering dominant modes of representation and cartography and stressing the ongoing colonialist and extractive logic inherent in that boundary. Philip E. Steinberg also cautions that the lines that divide ocean regions on contemporary legal maps of the sea "hide as much as they obscure," as the "history of the ocean is filled with attempts to mark off its spaces" in attempts either to claim territory or to create zones where certain activities by certain actors are permitted and others are prohibited.[40] To defy precisely this line-drawing logic, Steinberg emphasizes the materiality of the oceans, stating that they are to be understood as "physical entities"—"as wet, mobile, dynamic, deep, dark spaces" characterized "by complex movements and interdependencies of water molecules, minerals, and non-human biota as well as humans and their ships."[41] In an essay coauthored with Kimberley Peters, the authors propose to "explore the power of thinking through a wet ontology." They assert the necessity of going "beyond considering matter as static substance" and considering "the various ways in which matter changes physical state as it moves through, and simultaneously constructs, both space and time."[42] The fact that the ocean is not a metaphor, they argue, is precisely what creates the need for "new understandings of mapping and representing; living and knowing; governing and resisting."[43] In his insightful response to the article coauthored by Steinberg and Peters, Stefan Helmreich, an anthropologist of science, emphasizes what he terms a "killing ontology" by mobilizing their terminology to look at "the oceanic weave of transatlantic slave trade history and at today's long and deadly summers of migration in the Mediterranean."[44] In regard to refugees and, as mentioned before, their being continuously compared to a "surge" or "wave" that is flooding and threatening Europe, Helmreich points out that the "wave metaphor displaces attention from the real, deadly waves over which so many people have traveled" and that "waves that have become part of the dangerous *force field* that migrants confront as they cross the Mediterranean, often in overcrowded rubber rafts, fishing boats, and dinghies."[45] In propagating what we might call a matterphorical understanding of waves, Helmreich states that "water waves are forms that, with rescue boats struggling to do their work, pattern a deadly, fluid border," while all kinds of other waves—measuring, transmitting, and tracing signals—also play a major role in structuring the ocean and its force fields.[46] And in order to address the complexity of "ocean matter" and further investigate how materiality and metaphor relate across fields, disciplines, and matter(s), Jessica Lehman, Steinberg, and Elizabeth R. Johnson mobilize *turbulence* as a *matterphorical concept.*[47] In doing so, the authors sense and make-sense-with the concept's entanglements from three different entry points in order to "produce better engagements with the ocean's turbulent matter" by attending to "the lines and laws of the ocean," the "governance and epistemic cultures of ocean life," and "practices of marine historical knowledge production."[48] Concepts, to emphasize the importance of these interventions, *matterforth* and therefore require modes of sensing and sensemaking that are attentive to how matter and meaning are co-constituted. Given that 97.25 percent of the water on the surface of earth is found in oceans, and given that water is the most plentiful substance on earth and the third most plentiful molecule ($H_2O$) in the universe,[49] we might ask if legal theory can be forgetful or negligent of not only oxygen but also hydrogen. It is, after all, the most abundant element on earth and makes up over 90 percent of all atoms[50]—that is, three quarters of the universe's mass. How, then, does law relate to mass as opposed to land? How does it mobilize masses, criminalize or enable mass movements?[51] What makes its force massive? Are legal theories solid, sustaining political orders based on atomistic conceptions of how and what the world *is*? Do they require molecules to move as little as possible—or at least to make us believe that this is the case? Does it matter for legal theory that matter has become mortal? Can a nuclear bomb explode concepts, and how do they register radiation? Is law tied to specific states, compositions, and formations of matter? How tightly can legal theory hold particles, with or without mass? And what does any of it mean for alternative forms of relationality that desire to matter differently?

## Legal States of Matter

Approaching these questions will require extensive collaborative work and matterphorical case studies from various entry points. However, it can already be stated that legal concepts, despite many attempts to claim otherwise, are not solid. This becomes most evident in frontier spaces, where legal regimes or jurisdictions overlap, or when territorial boundaries are impossible to determine. At moments of uncertainty and indeterminacy, concepts, too, can crumble, dissolve, evaporate, or bind differently. The transformation from one state of matter into another has historically borne the potential to expose representationalist notions of legal concepts and to challenge, sometimes even dissolve, legal orders. For example, the transformation of the sky into sovereign airspace occurred due to the threat of planes dropping bombs from above state territories during the First World War. Similarly, the aerospace industry and the launch of the first satellite, Sputnik 1, in 1957 led to the development of a treaty system regulating activities in outer space.

Although airspace and outer space are prominent examples, they are not the only expressions that testify to how economic, physical, political, military, and legal forces co-constitute legal concepts and specific legal spatiality (and temporality). These forces not only shift lines and borders but also fundamentally change how various bodies exist in and experience space(time).[52] It is crucial to acknowledge that legal concepts matter and materialize; they are of the world and therefore sensitive to and expressive of temperature differences, molecular bonds, oscillations, vibrations, attractions, repulsions, and other forms of intimate onto-epistemological relationalities.[53]

Notwithstanding this fact, law, rooted in Western liberal thought and grounded in Cartesianism, relies heavily on representationalism, the imperialism of language, and the power of definition and classification. Law proper has been understood, as legal scholar David Delaney notes, as language: "interpretable text, as constellations of propositions, categories, syllogisms, rules, as linguistic expressions of thought."[54] It is imagined as stopping at the utterance: the verdict, the sentence, the decree.[55]

Imagined as a representational system, law, including its concepts, is thus considered to transcend its matter(s), thereby legitimizing its use, destruction, and exploitation. Both matter and materiality are seen as extrinsic, a foil on which law, independent from the physical world, has effects and leaves marks. This is also why the "dematerialization of the legal" must be seen for what it truly is: a "crucial boundary-making event" and, as such, always a political project.[56]

To demonstrate how matter, including its different states and constitutions, is onto-epistemologically co-constitutive of legal concepts and thereby entangled with political power and various forces at play, it is worthwhile to examine how international law governs the Arctic and the Antarctic. The Arctic and the Antarctic are not only different in terms of their materiality and physical states of matter, but these differences also play a major role in how they are legally determined and governed.

The Antarctic, for example, is regulated by the Antarctic Treaty System, established in 1959 after seven nations laid claim to territory in Antarctica.[57] The Antarctic Treaty System (ATS) declares that Antarctica "shall be used for peaceful purposes only," promotes free promulgation of scientific observations, and, perhaps most importantly, prohibits any claim *"to territorial sovereignty in Antarctica or create any rights of sovereignty in Antarctica."*[58] In addition, the Protocol on Environmental Protection to the Antarctic Treaty designates it "a natural reserve, devoted to peace and science" and ascribes to it an intrinsic value, "including its wilderness and aesthetic values and its value as an area for the conduct of scientific research, in particular research essential to understanding the global environment."[59] Based on that, the Protocol on Environmental

Protection states that "any activity relating to mineral resources, other than scientific research, shall be prohibited."[60] Despite the treaty system in place, Antarctica's legal status remains complex—discussions as to whether Antarctica is *terra nullius* (land owned by no one) or *res communis* (land collectively owned by humanity) or whether it falls under the more recent *Common Heritage of Mankind* principle (as applied to the seabed and the moon) are ongoing and unsettled. For the seven claimants, Antarctica is unequivocally owned territory. For nonclaimant states, it is, as a space where no state can make sovereignty claims, unowned.[61] Alexander Soucek speaks of Antarctica as a "continent 'in suspension'"; Christy Collis points out that Antarctica "comprises a composite of divergent spatialities" and is therefore not an unowned space but "in a state of geographical tension" as something between *terra nullius* and *terra communis*.[62]

In contrast to the Antarctic, the Arctic is mainly composed of the Arctic Ocean (rather than a land mass), is surrounded by various countries, and is covered with a perennial ice sheet that averages ten feet in thickness and can, in certain places, reach depths of forty feet.[63] What is more, the Arctic is populated. It is governed by a soft law regime, the Arctic Council, which serves as an advisory body and cannot issue binding law. Several treaties, for example, the United Nations Convention on the Law of the Sea (UNCLOS), the International Convention for the Prevention of Pollution from Ships, the Agreement on the Conservation of Polar Bears, and various other bilateral and multilateral agreements govern certain aspects of activity in the Arctic, yet they are silent on questions of sovereign control and environmental protection issues.[64] Because of its geographic location, with the Arctic Ocean directly abutting nation-state territories, the Arctic region includes territories of six sovereign nations (Canada, Denmark (Greenland), Iceland, Norway, Russia, and the United States), which exercise rights of coastal jurisdiction, including proclamation of 12-mile territorial seas and 24-mile contiguous zones and the functional expansion of jurisdiction to 200-mile exclusive economic zones and 250-mile continental shelf delimitations.[65] In addition to the climatic and the geographic characteristics, the geo-ontological status of the Arctic presents yet another challenge for legal theories that depend on analogies to land mass. It is not actually clear what it *is* that makes up the Arctic—an uncertainty, or indeterminacy, that lies in the fact that common means of measurement and categorization fail in the spaces considered the Arctic. Andrew J. Hund explains that, although commonly understood as the region surrounding the North Pole (90° N) which includes the Arctic Ocean, its marginal seas, and the coastal areas of the surrounding nation-states, the working definitions of the Arctic are related to *latitude*, *temperature*, and *tree line*. The definition oriented toward *latitude* defines the Arctic Circle as the area north of latitude 66°33′44″; this is closely tied to the conditions of light and darkness (as locations north of this latitude experience twenty-four-hour periods of sunlight on the summer solstice and twenty-four-hour periods of darkness during the winter solstice) and also moon gravity (which is responsible for the fluctuation in the position of the Arctic Circle). The definition based on *temperature*, at the ten-degree Celsius (fifty-degree Fahrenheit) isotherm, creates a shifting line, sometimes within the Arctic Circle (in terms of latitude) and often also outside it. Finally, the limit of *tree growth*, the biome boundary between the boreal forest and the tundra, is considered the southern edge of the Arctic and, naturally, shifts too.[66]

An even more concrete example for how law—in this case, international law—depends on physical states of matter and, consequently, the geophysical characteristics of spaces that hover in indeterminacy can be observed when taking a closer look into that which uniquely makes up huge portions of the Arctic and Antarctic: *ice*. The fact that Arctic sea ice, despite being not only juridically but also cartographically of the sea, also displays properties that more closely resemble land challenges the fundamental divide between land, transformable into territory, and water, defying territorial claims.[67] In looking at the physical and legal status of glacier ice, sea ice, shelf ice, icebergs, and ice islands in international law, Christopher C. Joyner argues that it is precisely their differing physical properties—ice being solid, water being liquid, both being different from land mass—that suggests the impossibility of applying legal norms by means of analogy: "Ice has solid physical properties, water is a liquid. The considerable distinctions between the two substances clearly suggest that similar legal norms cannot be neatly applied to both."[68] As a consequence, he calls for a "new international law" that attends to the "geophysical nature of ice and its legal relationship with the rest of the earth's environment."[69] Ice shelves, for example, surround more than one-third of Antarctica's coastline, account for 10 percent of the cold continent's area, and are legally not determined—neither in terms of the legal status of ice or in terms of a possible coastal baseline delimitation. Because, he argues, neither a land regime nor a high-sea approach is deemed a sufficient legal rendering—Antarctic ice shelves are considered neither frozen water structures nor analogs to land formations—shelf ice should be considered "legally as shelf ice, a unique kind of territorial space deserving unique legal treatment."[70] Similarly, ice islands, unique types of icebergs, "supply a new form of oceanic occupation that presents unique definitional and jurisdictional challenges for international law."[71] As approximately 75 percent of the earth's fresh water exists in the form of ice, with Antarctica holding around 90 percent of that volume—and as fresh water threatens to become scarce—polar ice will, scholars worry, increasingly become the target of resource extraction.[72]

Whether the solutions proposed by Joyner are a means to avoid growing international conflict over indeterminate legal spaces or a repetition of classical patterns in Western legal thought, it becomes clear that legal concepts *are* far from fixed and determinate representational units. They are *matterphorical* expressions, *mattering-forth* in material bodies and relations, that redirect matters of meaning by changing the dynamics of workability in a region.[73]

The onto-epistemological assumptions underlying law, be it national or international law, are certainly, to return to Delaney's argument, a political project: drawing boundaries, enacting exclusions, and making some modes of existence possible while excluding other from mattering. Challenging these assumptions and making different legal systems, even different understandings of law, thinkable and sense-able is thus an ethical imperative, a matter of *synaesethical* response-ability that requires fundamental shifts in both how law is imagined to exist and how this mode of existence makes (and breaks) sense on earth and beyond. Margaret Davies and Rhys Aston, for example, think law with and through the *ground* as matterphorical concept, asking whether we can speak "not only of human normative systems, but also bio-normative and geo-normative systems," and stating the importance of both "recognising and developing the relational and multi-agential sources and sites of human and more-than-human normativity," and "acknowledging and fostering their co-emergence and co-constitution."[74] The South African legal scholar and environmental lawyer Cormac Cullinan makes a related argument on a different scale when he states that legal systems are all based on the assumption that "we human beings exist only within our skins" and that "we are the only beings or subjects in the universe." Because laws are "generated entirely within our glass 'homosphere'" and thus "are understood literally, as laws unto themselves," the call for acknowledging a connection between the legal systems and the earth system goes unheeded.[75] For Cullinan, this is at the core of the current legal systems, which lead to exploitation—"dangerously unbalanced forces" based on property rights and the commodification of life and land—and injustices as already inherent in the concepts of law and jurisprudence.[76] To counter these injustices, Cullinan argues for an "Earth Jurisprudence" embedded within and simultaneously also extending the "Great Jurisprudence." The latter "is manifest in the universe"; in, for example, "the phenomenon of gravity [which] is expressed in the alignment of the planets, the growth of plants, and the cycle of night and day," and as such is "also written in the bones, muscles, sinews and thought patterns of our own bodies."[77] His argument is not that physical laws should determine positive, human-made laws but rather that the dissociation of legal theory, laws, and legal concepts from the world—its matter(s) and forces—is a political and philosophical strategy that creates the conditions for capitalist extraction and exploitation, biocide, mass extinction, and the destruction of ecosystems. In seeking appropriate legal forms (or concepts), Cullinan argues, we have to "look at the fundamental laws and principles of the universe," precisely because they provide "the ultimate framework within which any human legal framework must exist."[78] This is why Cullinan laments that knowledge about jurisprudence and law is still mainly acquired in "law libraries and lecture theatres from which nature is meticulously excluded," because "we are devising our legal philosophies and laws without reference to the 'primary texts' (i.e. nature) and seeking answers in libraries that do not contain those answers."[79] Earth Jurisprudence, however, "is not merely a theory, it must be a living practice, a way of life."[80]

Not only are law and physics entangled as fields or systems of knowledge production (epistemological entanglements) but the phenomena these systems aim to describe are also interconnected. My argument is not that physical laws and legal laws coincide, nor that they are directly applicable to each other. Rather, legal thought must become sensitive and *synaesethically* attend to the entangled phenomena that co-constitute laws and legal concepts. The task for *matterphorics* is to sense and make-sense-with the dynamism of intra-action or double articulation. This approach expresses an onto-epistemology of difference that co-constitutes laws and legal concepts rather than representing a legal world that neither exists nor is meant to exist. In addition, matter(s) of law and of physics are both confronted with indeterminacy because of their assumptions about the world, challenging rules, laws, and ideas about how the world is and should be. Near the beginning of his lectures on physics, attempting to relieve his students of totalizing ideas about scientific knowledge, Feynman states that all laws in physics are only approximations. He explains that the discovery of rules is considered "understanding" and claims that what is needed is both *imagination*, usually attributed to the theorist, and *experimentation*, which he describes as "the test of all knowledge" and "the *sole judge* of scientific 'truth.'"[81] However, he clarifies, "from the point of view of basic physics, the most interesting phenomena are of course in the *new* places, the places where the rules do not work," for it is in the new places that we discover new rules.[82] As we have seen in relation to frontier spaces and (epistemological and ontological) indeterminacy, this holds true for laws, legal systems, and modes of normativity too. The question to tackle through matterphorics is how to affirmatively relate to indeterminacy: how to response-ably and collaboratively construct legal concepts, plus modes of normativity and governance, with more-than-friends, guided by a synaesethics of thought. The aim is to think-with and create-from indeterminacy, moving beyond dualisms such as particle/wave, friend/enemy, and body/mind while remaining ethically sensitive to the limits of bodies and matter(s).

With regard to the onto-epistemological assumptions underlying classical notions of space and how those assumptions affect (human and nonhuman) bodies, the legal scholar Andreas Philippopoulos-Mihalopoulos declares his intention to "move beyond metaphors" and to respond to the "need to consider a material ontology on the level of the law which goes beyond (while not excluding) the textual comfort zone of the law."[83] Legal thought capable of *synaesethically* turning to law and legal concepts (accounting for the fact that they turn too) is attentive to force fields, dynamisms, and movements. It approaches a legal system as, to use Philippopoulos-Mihalopoulos's definition, "a snapshot of the differentiated velocity of its body, itself an assemblage that includes humans, technology, organic and inorganic matter as well as other social systems, taking place on a space of continuity with other bodies."[84] In doing so, its inquiry has to go *all the way down*—not leaving any assumption untouched, not hesitating to trace every concept to its point of indeterminacy. What happens to legal concepts, or what legal concepts can happen, if space, time, and matter are not separate entities? What if it is not legal spaces, times, and matters, but *spacetimemattering*—"a dynamic ongoing reconfiguring of a field of relationalities"—that legal thought must sense and become sensible to?[85] To return to Barad's quantum-field-theoretical interpretation, as well as to Deleuze and Guattari's plane of immanence, what if legal thought thinks-with a different ontology, an onto-epistemology of difference: one that does not allow for the assumption of the givenness, fixity, and separability of space, time, and matter; one in which matter "materializes and dynamically enfolds different spatialities and temporalities";[86] one in which phenomena are not located *in* space and *in* time but are *"material entanglements enfolded and threaded through the spacetimemattering of the universe"*;[87] and one in which sense is made *with* rather than *of* difference as onto-epistemological expression of existence? If these questions sound too intimidating or perhaps too abstract, we must remember that legal thought, as it stands, is built on Newtonian physics and very specific ontological assumptions. The question is: How can we think law differently—and would such a shift enable different modes of existence, relationalities, and expressions of justice?

# Rock Bottom

We do not know whether Goethe's wish was Schiller's command—whether Schiller decided to let the diver drown. According to the ballad's last verse, "the youth is brought back by no kindly wave."[88] Many readings are possible. Meaning does not have to be judged at the surface. Yet for what lies in the depths to matter and have meaning, it does not have to be unknown or hidden either. Although the ballad does not inform us of the diver's whereabouts, we do know that the deeper it gets, the higher the pressure rises. For some bodies, perception is severely challenged. Light rays entering and leaving water bend or refract because of their change in speed. Distances are perceived differently; suspended particles scatter and diffuse light, affecting the human eye's ability to see colors and forcing it to rely on contrast for visibility.[89] Urged to share what he had witnessed, the diver reported after his first descent (and ascent) that "under me lay it, still mountain-deep" and that "to the ear all seem then forever asleep."[90] Further from the coast, but also not "bottomless," the deepest point on earth measured under water is at 10,994 meters (36,070 feet) depth. It is deeper than Mount Everest, the highest mountain on earth, is high. And so who can blame the diver for his insufficient report on hydroacoustic matters, given that the first recording of sounds at the deepest points in the ocean— the Challenger Deep in the Pacific's Mariana Trench—was made only in 2015 using a hydrophone that could specifically withstand the pressure. Indeed, the maximum pressure recorded in the Deep was 11,161.4 decibars—a pressure equivalent to, as Olive Heffernan writes in *Nature*, "having a couple of elephants standing on your big toe."[91] The diver's statement that "the voice of mankind could not reach to mine ear" proves accurate. Yet the 2015 recording also revealed that anthropogenic, biological, and geophysical sounds, such as "earthquake acoustic signals (T phases), baleen and odontocete cetacean vocalizations, ship propeller sounds, air guns, active sonar, and the passing of a Category 4 typhoon" are audible.[92] Waves—propagations of disturbances from place to place, transferring energy—travel higher and deeper. The denser the medium, the faster they travel, because molecules are packed closer together, which facilitates the transmission of wave motion. Sound travels fast in cold sea water. Be that as it may, the king has not heard enough, and the diver might have reached his physical limits. The question of interpretation and of sensemaking, however, goes deeper. No matter how deep, Man's fatal attraction, it turns out, stoops low too. After returning from the first solo voyage to the deepest point in the ocean in 2012, having reached a depth of 35,787 feet, the filmmaker James Cameron heroically declared in an interview that the rationale behind his endeavor was the "exploration gene" and the urge to reach "the last great frontier on earth."[93] When the wealthy investor Victor L. Vescovo claimed only a few years later to have descended even deeper with his submersible, the question of interpretation took on another dimension. Cameron argued that one cannot go deeper than the deepest point, especially not when the surface in question is utterly flat. He contested Vescovo's record. As oceanographers explain, the uncertainty arises because sound waves are the main tool of undersea measurement; as mentioned, the use of sound waves as a measuring tool is complicated by the fact that the speed of sound varies with density, salinity, pressure, and temperature.[94] As of now, that matter is still unsettled, but we can safely say (perhaps with a margin of error in terms of the exact number) that the frontier myth runs deep.

That does not mean, however, that every*body* must bear the pressure of limits. The king, at least in this ballad, is not willing to risk his own life to gain knowledge about a space yet undiscovered, unresearched, and beyond his territory.[95] But things have changed. Now, rather than brave young men, robots are considered the greatest hope for reaching new frontiers. And yet, despite the contemporary ability to send remotely operated vehicles (ROVs) or autonomous underwater vehicles (AUVs) to register what lies in the depths of the ocean, and despite the technological advances in, among other fields, hydroacoustics and deep-sea spectroscopy, the greatest depths of the deep seabed have remained a space of scientific, financial, and legal uncertainty and indeterminacy, as well as of powerful curiosity. What is more, questions of sovereignty and property in the oceans are still, Elizabeth Havice and Anna Zalik argue, imprecise and subject to controversy.[96] Indeed, often described as the "last planetary frontier," deep-sea marine zones (as well as the high seas) are "facing growing pressures for extraction, resource-making and conservation," leading to "jurisdictional frontiers" and tensions that are "shaping and transforming notions of what jurisdiction is or should be in the oceans."[97] At the center of these changes are property rights, producing "massive change in the political economy of the oceans around neoliberal, market-based socio-environmental policies."[98] As the 2019 Greenpeace report on deep-sea mining "In Deep Water: The Emerging Threat of Deep-Sea Mining" states, the yet "nascent industry is ramping up to exert yet more pressure on marine life," thereby creating a "new industrial frontier in the largest ecosystem on Earth."[99] Because the International Seabed Authority (ISA), a 168-member body created by the United Nations to promote and regulate seabed mining beyond national jurisdictions, has already issued twenty-nine mining licenses for around one million square kilometers of the international seabed to sponsoring states working with corporate contractors, Greenpeace has limited trust in the ISA's ability to "protect the deep ocean from multiple other pressures."[100] Instead, Greenpeace calls for a strong global ocean treaty. The report "Why the Rush? Seabed Mining in the Pacific," released by the Deep Sea Mining Campaign, London Mining Network, and MiningWatch Canada, shares the mistrust in the ISA and its developing Mining Code, calling the Pacific Ocean the "new wild west" and requesting a twenty-year moratorium on deep-sea mining.[101] The report also states that, despite the fact that no mining has actually taken place yet, the deep sea has become a "speculative frontier," enabling investors to increase personal wealth yet causing governments, such as that of Papua New Guinea to take enormous losses that result in significant cuts in social welfare.[102]

While frontiers, although differing in their constitution and materiality, make indeterminacy sense-able, dominant modes of sensemaking, informed by profit-driven and extropist desires, appropriate and decide upon any possible determination practices. As Fountain writes, frontier spaces contain "large quantities of valuable natural resources" and environments "inherently inhospitable to humans." They require "technological sophistication and significant financial backing to exploit" and have "limited, if any, recognized sovereignty claims to date."[103] As Rasmussen and Lund argue, the appropriative logic and dynamics of the frontier have not disappeared. Rather, frontiers represent "the discovery or invention of new resources" and are therefore also the "sites of new forms of regulation and resistance and the emergence of new legal and social orders." This dynamic, as mentioned before, results in shifting terrains that move between legality and illegality and public and private ownership. More than a metaphor, "frontiers take place, literally," in spaces of potential danger and potential abundance—such as outer space, the Antarctic, the high seas, and the deep seabed—and are "replete with physical and symbolic violence."[104]

Frontiers, however, are not stable concepts either; they are material-discursive practices of boundary drawing, articulating im/possibilities for (modes of) non/existence. As such, the frontier, as a *matterphorical* dynamism, and the respective laws, legal orders, and concepts it co-constitutes (or prevents from coming into being) are neither comparable nor generalizable. Attention should be paid to their specific matterphorical expressions in order to intervene *synaesethically* into the frontier's modes of legal sensemaking. One such attempt has been presented by Eyal Weizman and his work on the "conflict shoreline," which denotes the "longest continuous aridity line on metrological world maps," beginning "in West Africa, just north of the Morocco/Western Sahara border."[105] The conflict shoreline indicates the "shifting threshold of the desert," connecting local histories but currently the site of conflicts all along its length. This boundary is determined not only by natural forces and conditions but also by an "interplay between meteoro-logical data (rainfall/temperature), patterns of human use (modern agricultural practices), and plant species (the cereal types used in intensive farming)." In examining the colonization of the Negev desert in particular, Weizman points out that despite being "one of the most contested frontiers in Palestine," the desert threshold is demarcated not by fences or walls but only on meteorological maps.[106] Importantly, he demonstrates, alongside photographer Fazal Sheikh, that this shifting aridity line has also been "an important political and juridical marker": "Decades after the state of Israel was established, a juridical mechanism based on the meteorological threshold of aridity was developed," marking everything beyond that line as unsuitable for permanent settlement and agricultural cultivation. This, in turn, was used to justify expropriations and denial of Palestinian property rights.[107] However, Weizman and Sheikh note that with temperature and evaporation rates on the rise, a "formidable counter force" has arrived in the last decade, pushing the desert in the opposite direction, aggravating existing conflicts, and engendering new laws and legal constructs.[108]

The discovery of "new places" (colonized territories, the high seas, Antarctica, the deep seabed, outer space, cyberspace, and crypto-space) has led to the creation of new rules and significant legal shifts. These changes, which often undermine protective laws and lead to lobbying for more lenient regulation, are too often driven by colonial, extractivist, and expansionist powers. However, it is a misconception to assume that indeterminacy, shifts in our understanding of the world, or technology will inevitably and exclusively foster structures of oppression, domination, and absolute determination.

Matterphorical legal case studies reveal an urgent need for new concepts—ones sensed-with the matter(s) at stake, arising from points of indeterminacy where matter and meaning are co-constituted and that matter-*forth* responseably. This practice of doing theory, I argue, is critical yet does not stop at critique. It synaesethically creates alternatives for and with more-than-friends that enable different modes of existence. Importantly, matterphorics has neither a center nor an aim to establish rules and regulations. Rather it is a decentralized mode of sensemaking that values multiplicity, diversity, and difference—embracing a thousand approaches, thereby avoiding centers of power and determinism. Matterphorics as analytics of existence is a method for critical diagramming, mapping force fields, tracing power and potential. As a practice, matterphorics is a tool for traitors to the world of dominant significations and the established order, who do not deny complicity. Quite the opposite—they engage with frontier spaces, where potential attracts, falls entice, and newness tempts, yet do not become enamored with power. It is precisely in frontier spaces where traitors, synaesethic thinkers, are most needed—to build with more-than-friends and oddkins the unimaginable and unthinkable, alternative and more livable concepts, but also to ensure access and facilitate knowledge transfer for those consistently kept out, for more-than-friends, allies of the common (yet not the same). It is then that we break the image of thought and thus also the unwritten law described by international legal scholar Martti Koskenniemi: "concepts and structures that are themselves indeterminate nonetheless still end up always on the side of the status quo."[109]

This also means, however, that synaesethics does not imply avoiding power. Instead, as Massumi states, it demands creatively getting "down and dirty in the field of play," mobilizing complicity toward new kinds of emergences, staying with the trouble, and adamantly refusing to believe that "the game is over" or that there is "no sense having any trust in each other in working and playing for a resurgent world."[110] Yet trust must be forged not in the shadow of a common foe nor under the dominion of Man. This trust demands time, for the image of thought still reigns, favoring sameness and denying difference its legitimacy. As this section shows, it will be no easy task—especially as Man seems to rise perpetually, even when He appears to have finally fallen, only to fall again: higher, faster, fiercer.

**III**

# Cutting-Edge Theory
*(Life) Story Telling*

# 10

# Matterphorics of Life

## AN ALL-TOO-REAL STORY

In the trees
between the leaves
all the growing
that we did

all the loving
and separating
all the turning
to face each other

I divide
in the sky
in the seams
between the beams
—J. RALPH AND ANOHNI, "Manta Ray"

To open this last part, let me return to the beginning of the first: to Haraway's dismay about Latour's recent turn to Carl Schmitt, not only regarding the question about whom to think-with but also regarding how *life* is constructed. In the seventh lecture of his *Facing Gaia*, Latour claims that because of disputes over climate and how to govern it, the "political question" is again raised "in terms of life and death," urging us to ask, "in the name of what supreme authority have we agreed to give our lives" and to take—even sacrifice—those of others.[1] Given his articulated concern, Latour's conclusion is rather surprising. With confidence, he suggests returning to "the toxic and nevertheless indispensable Carl Schmitt," the Nazi legal scholar that he maintains "can be likened to a poison kept in a laboratory for the moment when one needs an active principle powerful enough to counterbalance other even more dangerous poisons: it is all a matter of dosage! In the case in point, the drugs we have to counter are so strong that I invite you to desensitize yourselves with small doses of Schmitt, taken advisedly."[2]

Latour's *comparison* of Schmitt's concepts, which originated from and were designed to aid Nazi ideology, to a poison that can be kept isolated in a laboratory necessarily fails, and with it the metaphor he constructs.[3] Thought (and concepts) cannot be put under quarantine and simply distributed as needed by those who claim to be in control of its effects. What's more, as Haraway writes, matters of thought and storytelling are "immensely material." Indeed, she not only states that "it matters what stories make worlds, what worlds make stories," but also emphasizes that it matters "what stories we tell to tell other stories with."[4] Her caution reads then as follows: "Schmitt's enemies do not allow the story to change in its *marrow*; the Earthbound need a more tentacular, *less binary life story*. Latour's Gaia stories deserve better companions in storytelling than Schmitt. The question of whom to think-with is *immensely material*."[5]

In this part, I wish to think precisely about this *life story*: how binaries are upheld, what the role of law is in creating and sustaining this story, what it has to do with genre, what concept of life underlies it, why it matters, and what theory can do. Our focus lies on the complicity between not only life and legal theory—what it means to have a right to (*a*) life—but also literary theory. Haraway's criticism of Latour comes in recourse to the trope of the Anthropocene as earth's life story, or, put differently, the story about *life* on earth. Who will have the right to *a* life?[6] Which kind of lives are protected? Which species will be resurrected, and which will become extinct? Whose habitats will be destroyed, whose water taken, to sustain the lives of whom? These questions circulate within the context of the Anthropocene, a concept that has significantly shaped modernist conceptions of life, temporality, and spatiality. However, the Anthropocene fails to offer a less binary life story, instead reproducing an all-too-familiar narrative within an all-too-familiar genre. In fact, it is helpful to understand the Anthropocene through a matterphorical understanding of what Derrida called the "auto-biography of man." It finds its legitimation in Enlightenment conceptions of the liberal subject, transcendence, individualism, human exceptionalism, and anthropocentrism—all of which, as literary scholars have pointed out, are characteristics of the traditional autobiographical genre. As a life story, the Anthropocene doomily proclaims Man's fall, relentlessly proclaiming that it is in earth's best interest to make him rise again, this time more sustainable and green. Falls, we have seen, are a matter of forces. "Unless there is a global catastrophe—a meteorite impact, a world war or a pandemic—mankind will remain a major environmental force for many millennia," Paul J. Crutzen famously stated in 2002 when he introduced the term *Anthropocene* to a broader audience.[7] As the sociologist and science and technology studies scholar Eileen Crist writes, this "Promethean self-portrait" delivers the notion of "an ingenious if unruly species, distinguishing itself from the background of merely living life, rising so as to earn itself a separate name," and celebrating its "unstoppable and in many ways glorious history (created in good measure)," which has "yielded an 'I' on a par with Nature's own tremendous forces."[8] As such, the Anthropocene marks the geologically defined moment when Man, as a particular form of human existence, is declared a force powerful enough to "overwhelm all other biological, geological, and meteorological forms and forces."[9] Although there still is no consensus on when exactly this moment should have taken place, the geological time scale makes it possible to structure an unimaginably long time (the age of the earth is said to be about 4.5 billion years) in intervals based on major earth-historical events, including the appearance or extinction of significant life forms registered by rock layers—sometimes continuous, sometimes severely interrupted. Despite the progression of human-induced global warming, its cascading ecological, political, and social effects, and the imperative for a collective response that preserves the planet and builds more equitable and sustainable systems, the Anthropocene, following the Pleistocene and Holocene, is not a story about existences but one about an anthropocentric, racialized, and gendered concept of *life*. As Michael J. Moore argues, the Anthropocene's historical perspectives can be narrowed down to two narrative strategies, both rooted in Cartesian dualism: first, that the periodization is determined by consequences, and second, that Anthropos is said to engender these consequences. Moore states that "the Anthropocene makes for an easy story" precisely because it reproduces oppressive and exclusive structures, "doesn't challenge the naturalized inequalities, alienation, and violence inscribed in modernity's strategic relations of power and production," and "reduces the mosaic of human activity in the web of life to an abstract, homogenous humanity."[10] Povinelli demonstrates how the Anthropocene's narrative—according to which "the end of humans excites an anxiety about the end of Life and the end of Life excites an anxiety about the transformation of the blue orb into the red planet, Earth becoming Mars, unless Mars ends up having life"[11]—entails a cut, that is, a separation of life from nonlife.

Who or what, then, can write this all-but-fictional story, a narrative written by, among other material-discursive practices, CRISPR-Cas9 rewriting DNA, genetic engineering altering soy seeds, solar-radiation management modifying the atmosphere, AI models governing ecosystems? As it is, this story, Haraway attests, has only one actor: "all others in the prick tale are props, ground, plot space, or prey. They simply don't matter: their job is to be in the way, to be overcome."[12] How do we decentralize authorship and storytelling? How do we create the conditions for a thousand different stories—mindfully embodied and matterphorical—that enable liveability for more-than-friends and, importantly, challenge the Cartesian cut that continues to cut matter(s) of existence into life and nonlife, rending the latter inferior, property, tradeable, disposable, or killable?

Within legal theory, the Anthropocene concept's shortcomings have been pointed out and its disruptive potential has been appreciated and mobilized. Important calls for shifts in legal thought and practice toward "addressing more fundamental questions about how law shapes the possibilities and conditions of life" rather than simply "saving the planet" have been articulated.[13] However, this is not an easy task. Legal thought heavily relies on the separation between law and nonlife, most significantly articulated in the distinction between legal subjects and property. This distinction affords the subject not only the right to (a) life but also the right to freely dispose of other lives, life forms, and modes of existence. Influenced by the rationalist strand of Western philosophy, it determines subject-object relations, which, as legal theorist Anna Grear argues, are fundamental to the genesis of the Anthropos and crucial for understanding the exclusions enacted by anthropocentric law.[14] This distinction dictates what kind of matter simply does not matter, leveraging ever new tools to do so. Scientific and technological advancements, guided by the same logic, have produced intricate entanglements between life and property. As Stefan Helmreich observes, "biotechnology, biodiversity, bioprospecting, biosecurity, biotransfer, and other things *bio-* draw novel lines of property and protection around organisms and their elements (e.g. genes, organs)," which "now circulate in new ways as gifts, commodities, and tokens of social belonging or exclusion."[15] These novel property lines intersect closely with issues of copyright and patent law. Sheila Jasanoff compellingly demonstrates that the law governing life patents is "logically consistent with the law of 'takings' as it relates to real property."[16] The 1980 Supreme Court decision in *Diamond v. Chakrabarty*, famously stating that "the fact that micro-organisms are alive is without legal significance for the purpose of patent law," granted a patent for a genetically engineered bacterium.[17] This opened the door to "patenting any living things that were the creation of human hands and human ingenuity," extending to many higher animals.[18] Jasanoff shows that granting patents on living things involves "decisions about where to draw the line between life and matter."[19] Kathryn Garforth, in turn, emphasizes the conception of patents as "a series of fences that delineate the invention from other patents and the public domain," conveying the impression that "the inventor has control over the invention."[20] What becomes clear is that patent law is a powerful legal tool to territorialize life. As Vandana Shiva argues, it moves power from landlords to lifelords, akin to how the jurisprudence of *terra nullius* allows mass colonization by defining land as empty: "The jurisprudence of intellectual property rights related to life-forms is *bio nullius*: life as empty of intelligence. The Earth is defined as dead matter, so it cannot create."[21]

As demonstrated, the power to determine modes of existence based on likeness and difference is a fundamental characteristic of the representational modes of thinking challenged throughout this book. Along with capitalist logic, this is precisely what allows North American courts to eagerly conceive of living organisms as akin to chemicals, reducing them to chemical descriptions.[22] Similarly, arguments in favor of extending copyright protection to life forms and engineered genetic sequences have been driven by analogies between DNA and software code.[23] The pervasive influence of representational and proprietary thought extends across scales. Widely transmitted analogies between life and text or machines, for example, strongly impact bioengineering practices and the legal and economic regimes that are installed in and around them.[24] As science historian Sophia Roosth argues, the presence of both copyright and patents in synthetic biology "is enabled by assumptions of what life *is like* as well as how it *should be* remade, debugged, rewired, or rewritten."[25] Read together with Jasanoff's attestation that law "constructs both life and capital,"[26] it becomes clear how law and the binary life story relate. Also clear is how (and why) subject-object relations are such powerful instruments in determining what life, as a concept, is for/in/through law and legal thought. What all these examples reveal is that the narrative of the Anthropocene as a geological age is upheld by a particular relation to *life*, fostered by proprietary legal thought and practice and enforced by molecular cuts and lines.

Within legal theory, these developments and their underlying assumptions about life have been critiqued in ways that emphasize the necessity of investigating how law and matter relate when it comes to the very concept of life. In shifting the terminology from Anthropocene to *Nomocene*, Delaney exposes the fact that there is no life-as-singularity that does not affect, or is not affected by, what we call "law," and that human and nonhuman forms of life are all inextricably woven into the web of law.[27] Relatedly, Andreas Philippopoulos-Mihalopoulos rethinks law by building on immanence thinkers and feminist neomaterialists, arguing that there can be no clear-cut boundary between human and nonhuman bodies and their environment.[28] In a similar vein, Cullinan states that laws have been generated solely within our "glass 'homosphere'"—articulated and conceptualized by human language, driven by the false belief that there is no need for any connection or continuity between the legal and the earth system.[29] Matters of law, as demonstrated in the previous part, are indeed immensely material. Consequently, we must be attentive to how legal theory encounters and enables stories of life and how life stories in turn inform and enable legal theory. Indeed, the complex entanglement of *life* with legal and literary theory requires becoming aware of how *life* is pressed into forms, stories, and genres and of how law is complicit in upholding the binary life story Haraway urges us to renounce. What underlies this part of the book, leading to its final matterphorical case study, is a questioning of the binary that introduces life through its negation (nonlife), rendering various forms of being and existence (human and nonhuman, animate and inanimate, vibrate and inert objects) external to what is held up and considered *a* life. My interest here resonates with Povinelli's claim that "it is certainly the case that the statement 'clearly, x humans are more important than y rocks' continues to be made, persuade, stop political discourse," making comparison and absolute difference (or the negation of sameness) the exclusive tools of determining where the questioning ought to end. The final chapter in this part speaks to Povinelli's attention to "the slight hesitation, the pause, the intake of breath that now can interrupt an immediate assent."[30] What if this breath carries the potential to radically rethink what a right can be and do?

# When Theory Crashes
# (into) Life

What follows is the greatest story ever. It's the story of... "the genome." Its being—its existence across time, its depth and complexity as a natural artifact, and the vast abundance and variety of its manifestations—is the story. —CHURCH AND REGIS, *Regenesis*

Later on we discovered that a single "spelling mistake"—the deletion of just one base—out of 1.1 million letters of genetic code meant the difference between life and death, when it came to creating the first synthetic cell. —CRAIG VENTER, *Life at the Speed of Light*

Every metaphor breaks down somewhere. To **have** a story, and to **be** one, are not the same. —GEORGE ESTREICH, *Fables and Futures*

Before examining how law divides entangled modes of existence into *life* and nonlife and the narrative strategies that underlie these cuts, I wish to look more closely into how life and theory encounter each other and the consequences and perils therein. Although I have not explicitly focused on *life* in the previous chapters, I have argued that doing theory *matterphorically* allows us to articulate ontologies of difference and relationality, which is the prerequisite for a *synaesethics* of thought and for post-anthropocentric knowledge production. *Life* has inhabited the theories, matterphorical case studies, and even the matterphorical *con*figurations I have used throughout the chapters—among them the brittle star, legal force, and the plane of immanence (or unique plane of life). Deleuze, for example, titles his last theoretical text "Immanence: A Life."[1] Barad, as previously mentioned, argues that "all life forms do theory."[2] Braidotti describes thinking as "life lived at the highest possible power,"[3] Whitehead states that the function of reason is *"to promote the art of life,"*[4] and Haraway's "Situated Knowledges" calls for critical theories that can "build meanings and bodies that have a chance for life."[5] While in each case we might get a sense of what *life* might *mean*, what is understood as *life* in these instances is neither identical nor comparable. What's more, none of these mentions of *life* offers, or attempts to offer, a clear and universal definition of what life *is*. The fact that life reappears throughout this part, however, is significant; it reminds us that an ethics of doing theory matterphorically touches, at one point or another, upon the question and matter of *life*—understood as neither purely conceptual nor inert biological matter deprived of social, legal, and political meaning.

I have shown in the previous chapters—in the metaphorization of the sun and the equalization of missives with missiles and rhetorical force with nuclear force—how metaphor, as that which conveys meaning from one conceptual realm to another, is not merely a creative tool of meaning production. My attention was mainly directed at philosophy, as well as literary and legal theory. However, metaphor, and the critique thereof, plays a major role in the sciences too. Indeed, the sciences have a long history of using metaphors in particular, and analogical reasoning in general, to create new concepts, productively invent meanings, communicate findings, and creatively fill the gap that arises between language and, depending on the scientific tools, observation, experiments, or equations.[6] In fact, metaphor has even been considered as "that which makes theory possible," a way of "connecting different orders of reality," and as that which provides for a "referential imprecision [that] can have a positive function in scientific work."[7] While they certainly bear the potential to lend meaning to what is inexplicable or difficult to convey in representational terms, metaphor and analogical reasoning—when set too free from their material-discursive fields—have lent themselves to misuse, appropriation, and misinformation. This is not a new claim, but it has been revived with the rise of genomics and its lead metaphor of the "book of life" and with the development of synthetic biology; the biologist is assimilated to an author of life-forms, who rearranges biological components like Lego bricks to "rewrite" this book.[8] Etymologically, metaphor denotes the "carrying over" of meaning from one word (or concept) to another, usually by means of analogy. When it comes to *life*, metaphor is challenged most directly because life, whether imagined as singularity or as everything living on earth (and beyond), is specifically not—not even conceptually—transferable (except for transhumanists like Moravec or More). In an article called "A Synthetic Creation Story," the science writer for *Nature*, Philip Ball, debunks the metaphor of genome sequencing as the reading of the "book of life" and of synthetic biologists as authors of life. He states that life is not "a thing one makes, nor is it even a process that arises or is set in motion," but "a property we may choose to bestow on certain organizations of matter."[9] Speaking from the perspective of the natural sciences, he emphasizes the problematic metaphorical extension of life across space-times: "'life' in biology, rather like 'force' in physics, is a term carried over from a time when scientists thought quite differently, when it served as a makeshift bridge over the inexplicable."[10] This invokes the gap between matter and meaning, which is critical when it comes to concepts such as (a) life or the right to (a) life. It also underscores the problem that a mode of doing theory ethically and matterphorically cannot simply remain in a conceptual space where meanings are transferred, altered, and distorted at will. That the encounter between theory and life is, regardless of discipline and method, highly complex and rattled with uncertainties, incommensurabilities, and indeterminacies that *matter*, is revealed by a glimpse into various contemporary attempts to address this problem.[11]

Biology offers itself as a convenient entry point, given that it literally means the "study of life." Does biology have a *theory of life*, and if so, what does *life* mean in the field of biology? In *Making Sense of Life*, the science historian Evelyn Fox Keller points to the fact that in biology there is a persistent uncertainty that complicates more than just the question of what life *is* and *means*. It also inhabits ambiguity concerning life's origin, as it refers to both the emergence of life on earth and the unfolding of a singular (or individual) life, with both forms often being intertwined and overlapping.[12] Life, as a category, oscillates between "being located at the level of (at least) the gene and emergent at the level of the globe."[13] Although biology's main task is generally understood as to "make sense of life," and despite the discipline's successes over centuries in describing, analyzing, and observing what has been termed *life* at respective times in history, biology, Keller argues, has neither a "theory of life" nor a shared "concept of theory." In fact, unlike physics, for example, which strives for a "theory of everything," biology has, throughout most of the past century, "generally eschewed the possibility, or even the value, of an overarching theory of life."[14] In the end, Keller states, biology "is scarcely any closer to a united understanding (or theory) of the nature of life today than it was a hundred years ago."[15] This is all the more interesting as the rise of genomics and synthetic biology over the twentieth century has significantly influenced how questions of life, and of whether or not there can be a theory of life, are asked and addressed. In her book *Can Science Make Sense of Life?*, published in 2019, Sheila Jasanoff expresses her worries not only about the entanglements of science and biocapitalism but also about how, as a consequence of the great breakthroughs in the life sciences in the twentieth century—precisely genomics and synthetic biology—it has become "increasingly more acceptable for biologists to claim ownership of the meaning of life."[16] As intellectual property rights over life-forms and living organisms show, ownership of the meaning of life is inextricable from that of pieces of biological matter—in some cases considered (a) life, in others simply rendered property.

The anthropologist and philosopher Elizabeth Povinelli addresses the question of *life* from a different perspective. For her, life (the concept of life) functions through its opposition to nonlife, constructing a form of power (*geontopower*) that cannot be fully described by Foucault's concept of biopower and is most intensively present in areas of late settler colonialism. Critical theory, in her opinion, is increasingly incapable of demonstrating the superiority of humans to other forms of life or of maintaining "a difference that makes a difference between all forms of Life and the category of Nonlife."[17] This has also to do, she argues, with the concept of the Anthropocene, which has exposed the entanglement of geology with life and discredited the long-maintained distinction between the biosciences and geosciences, making it increasingly difficult to uphold a notion of life as distinct from nonlife (*geos*), let alone the unicity of any particular life-form.[18] Here too, the question of life returns to that of entanglement (*extimacy*). It shows not only the untenability of a separation between life on earth in general and life individualized and localized in its tiniest bits; it also points to the fact that, although "to be 'life' a living thing must be structurally and functionally compartmentalized from its environment," nothing, Povinelli argues, "can remain alive if it is hermetically sealed off from its environment."[19] This is also why critical theory has, she argues "increasingly put pressure on the ontological distinctions among biological, geological, and meteorological existents" and why "posthuman critique is giving way to a post-life critique."[20] From a geological point of view, then, a *theory of nonlife*—of its material and conceptual properties—might be an appropriate starting point into theorizing *life*: one capable of accounting for *geontologies* (rather than biopower) and therefore liberal governance in and through late settler colonialism.[21] The complexity of the issues at stake requires a different theory, as well as a different mode of thinking-with. It demands not only a postlife critique, or a postlife critical theory, but also a "new interdisciplinary literacy" to find a way to "square our current arrangement of life with the continuation of human and planetary life as such." The question for Povinelli is less one of unbridgeable gaps between methods, languages, and practices and more about common frameworks and the "unacknowledged agreements [that] were signed long before the natural and critical sciences parted ways."[22]

Kimberly TallBear, an indigenous studies and science and technology studies scholar, looks at precisely these tacit agreements when addressing the ethics of cryopreservation of indigenous biological samples. She points out that it is particular notions of *life*, namely "nonindigenous binary concepts of life versus death and human versus nonhuman," that underly the narrative of the preservation of indigenous lives and indigenous bodies and even of the relations between scientists and indigenous peoples.[23] This particular narrative, she writes, calls for "preserving remnants of human groups and their nonhuman relations, defined in molecular terms, and archiving those molecular patterns and instructions" precisely because "indigenous death is far enough along to justify appropriating indigenous resources." It supports a "genomic rearticulation of indigenous life as the rightful patrimony of global society."[24] TallBear demonstrates that the narratives told by "genomic story-tellers" mobilize a concept of *life* that utterly depends on Western scientific understandings of history and progress. Speaking as an indigenous scholar, she therefore cautions against allowing old divides like the life/not-life divide—which are upheld by institutions to govern indigenous lives, the land, and the lives of nonhumans who have been savaged by Western analytical frameworks—to pervert understanding of indigenous lifeways.[25] This is yet another way of pointing to the *marrow*, even the molecular structure, of a binary and universalized life story that continues to cast just one actor.

Sophia Roosth and Stefan Helmreich expose, in different ways, how at the limits of concept and theory, *life*'s boundaries begin to blur too. Dispelling the assumption that "theorizing life always relies on animation and vitality," Roosth tells the "history of life suspended," that is, of organisms that challenge our assumptions about life precisely because, in periods of suspension, they simply do not grow, metabolize, move, perceive, or respond to the environment.[26] And she does so by focusing on a series of biomedia, "each of which incarnates a particular liminal vitality and announces a specific temporal problem inflecting theories of life."[27] From the static temporality of petrified biological media (fossils) to the pausable temporality of seeds and blossoms that can remain dormant for sometimes even millennia—life in ice or crystal and life dried out into dust and ash, displaying forms of suspended life with a reversible and discontinuous temporality—Roosth demonstrates how what inhabits (quite literally) the deepest crevasse of metaphysical divides, constantly hovering between life and nonlife, *is* what theorizes life.[28] Theory, here, relies on *matterphorics* as life in each biomedium "reticulates into multiple materialized theories of itself."[29] The petrified, frozen, dried, powdered, irradiated, and dormant life-forms are "material apparatuses manifesting theories of what life is, what it is not, and what lies in between," implying that theory and matter are co-constitutive.[30] The urgency of tracing the limits of what is considered life and a life-form, as well as of narrating a history of that which does not take the separation between life and nonlife as a given, evolves from recent developments in the life sciences. Synthetic biologists, for example, create the meaning of life—what life *is*—by manufacturing biotic media, new living entities that are, Roosth states, "theories, materialized."[31] In other words, in the postgenomic era, life is not the common dominator of all living things but has become once again "a problem of ontological limits and discontinuities."[32] Certainly, the discontinuities at the core of life's temporality require close attention as they unsettle linear forms of life narration, construed not only biologically but also socially, politically, and legally. In *Synthetic Life*, Roosth puts the relation between life and narration most directly: Synthetic biologists, as both "agents and participants in a grand evolutionary narrative," construct new life-forms as an attempt to "retrospectively define what counts as 'life' to accord with the living things they manufacture and account to be living."[33] However, despite the binarity at the core of the hegemonic life story Haraway criticizes, the ontological limits and discontinuities revealed by life's complex temporalities will again, if not properly addressed, be forced into the ever-same extractive and appropriative narrative. Biotechnologies such as somatic cell nuclear transfer (SCNT) and their potential for the resurrection of selected extinct species are effectively reversing biological time.[34] In addition, the rise of private cryonic facilities, such as the Alcor Life Extension Foundation founded by extropist and transhumanist Max More, as well as the costs of storing the dead bodies of the "immortalists" who can financially afford to believe in transcending death, demonstrate—especially in relation to TallBear's arguments about cryopreserving indigenous DNA—how chosen future lives and life-forms are inscribed in the life story's chapters to come.

Helmreich, too, approaches life and theory from their respective limits. In the preface of *Sounding the Limits of Life*, he writes that the "conceptual trouble" befalling "life," specifically one that is "indexed by worries about what constitutes its essential 'form,'" is "shadowed by worries about what form 'theory' might take in natural and social analysis these days." Theory, he argues, is pushed to its limits by a world no longer satisfyingly captured by "theoretical scaffolds" such as nature, culture, and society "inherited from mostly European nineteenth-century sources."[35] Now increasingly referred to in quotation marks denoting a dissensus about its meaning, *life* becomes "a shadow of the biological and social theories meant to capture it."[36] Looking at limit biologies, such as the transhumanist conception of artificial life, extreme marine microbiology, and astrobiology, exposes the "growing instabilities in concepts of nature," as well as the fact that "there is no once-and-for-all theoretical grounding for life."[37] Helmreich's approach, then, articulates a matterphorical take on *life* as "amalgam of the conceptual and the actual," a "substance-concept," ignoring neither the material nor the discursive practices that shape the concept.[38]

What these contemporary approaches to how life and theory relate share with each other is their attentiveness to the *matterphorics* of life: to the way meaning and matter, in their many compositions, are inextricably entangled. It is not only physics that needs to face the fact that our entire picture—or, to remain within the context of the binary life story, our narrative(s) of the world—has to be altered "even though the mass changes only by a little bit."[39] Put differently, there is no such thing as a transfer of meaning (of a concept or a word) without matter; neither is there a change in matter that does not also elicit a shift in meaning. Neither theory nor concepts can simply be applied. This becomes especially apparent when seen in relation to *life*. For what makes it a contested concept, a semantically unstable word, a shifting category, an unsettled matter, and a continuous concern is that *life* can hardly be understood metaphorically. Every cut separating life from nonlife (the inert, the dead, the not-yet-alive) not only *matters* but is most legible and most intensely sensible to that which is closely entangled with it. While *life* might easily be turned into a metaphor from afar, it is all too real when it is negotiated, threatened, and redefined in close proximity to the bodies, matters, or places from where its meaning is called into question. It is not a novel claim that *life* is, perhaps even more so than "force," (an) unsettled matter, mindful and mindless, to degrees not measurable and not fully conceptualizable by human modes of thought and practice. Definitions of *life* are highly consequential cuts through matter and concepts. These onto-epistemological cuts are what creates *life* and what renders the remainder, what is cut off, insignificant. Looking into how normative frames produce recognizable lives, Butler draws attention to precisely this dynamic and to the uncanny overflow (or "remainder") of life by explaining how normative frames "work to differentiate the lives we can apprehend from those we cannot"; that is, how they "produce lives across a continuum of life."[40] Although *a life* "is produced according to the norms by which life is recognized," there is "a remainder of 'life'—suspended and spectral—that limns and haunts every normative instance of life." Its production can only ever be partial, remaining "perpetually haunted by its ontologically uncertain double."[41]

This is precisely why the relation between legal theory and life requires careful attention to both material and discursive practices. In Western legal systems, the legal device that serves as the frame for recognized lives is that of the *legal person* (or subject) and the device that frames the "remainder" is *property* (or object). This construction is problematic, to say the least. As Delaney summarizes, law is "largely indifferent to how life is subordinated through property to the will of the minded legal subject."[42] Legal subjectivity and property are the "the two major concepts employed by law to classify the world." As a consequence, the distinction between them is "one of the most fundamental divisions of legal matter."[43] Stated differently, the reduction of nonhuman life-forms (let alone nonlife) to thinghood is followed by their legitimated subordination to the "desires and fears of particular possessive subjects also referred to as 'owners,'" who therefore have their things and objects at their free disposal.[44] The distinction between legal subject/person and property, although often defined as mutually exclusive, is similarly intricate, precisely because in liberal philosophy the person is both "subject and object of her own property, existing as a self-relation which is divided and yet whole, for instance as (owning) mind and (owned) body."[45] Although the broadly accepted legal view that there can be no property in humans suggests a rather clear distinction between human lives and property, biotechnology, life patents, and cases like *Re A* (see chapter 12) show that the Cartesian cut operates according to not only a neohumanist but also a biocapitalist logic. Aided by science and technology, law can now, the legal scholar Ngaire Naffine writes, "permit or prohibit some kinds of being from coming into existence" precisely because, as Jasanoff argues, it "demarcates those aspects of life that can be owned from those that cannot."[46] The fact that new biologies cross the dividing lines between "life and nonlife, human and nonhuman, individual and collective, predictable and unpredictable"—which have long been foundational to legal theory and thought[47]—exposes the onto-epistemological cuts at the core of Western thought, far exceeding singular disciplines and their respective abilities to assess the extent of these cuts.

To invoke literary theory, the modes of narration that underlie both the binary life story, as well as the story of *life* as a concept upon which legal, political, scientific, and social discourses rely, reveal their complicity with legal theory in shaping these stories and preventing others from taking hold. Patents on life forms such as bacteria, seeds, or the OncoMouse and copyright inscriptions into the genome of synthetic microbes, as performed by scientists at the J. Craig Venter Institute (JCVI), are cases in point. Intellectual property rights are inextricable from literary history and theory, as they share the concept of an ingenious author and creator. The author, understood "in Romantic terms as an autonomous individual who creates fictions with an imagination free of all constraint" and for whom "everything in the world must be made available and accessible as an 'idea' that can be transformed into his 'expression' which thus becomes his 'work,'" links intellectual property rights with a certain construction of (literary) authorship.[48] This liaison, however, is far from inconsequential. The attempt to "clothe a newly invented Romantic author in robes of juridical protection" has wide-ranging negative consequences, "with costs in areas ranging from biodiversity and the production of new drugs to the shape of the international economy and the structure of the computer industry."[49] In a similar vein, Martha Woodmansee and Peter Jasz emphasize the "possessive individualism" underlying this concept of authorship and argue that the prevailing construction of the author "as the bearer of special legal rights and cultural privileges" entails "consequences for the ways in which power and wealth are distributed."[50]

While I will focus next on the complicity between theories of life, law, and life writing, the entanglement just described already shows that this is a question not only of what theory can (actually) do but also of how it encounters *life* matterphorically. This is crucial because, as with other encounters, modes of theory engaging with *life* can fall prey to precisely the proprietary and exclusive logic that results from the classical image of thought (part I) and modernist legal concepts, especially when paired with anarchocapitalist desires (part II). What does it mean for a *theory of life*, or for its material bits and conceptual parts, to be considered the creation or invention of an ingenious individual, an author that can legitimately claim a legal right to it? What do we do with the fact that, as I will explain later, the epistemological, normative, and ontological cuts performed on matter(s) of existence are deemed inherently different based on the scale on which they are perceived? Where exactly is this unquestioned line (or cut) that presumably severs theory from life—and which disciplines decide that, according to which modes of thought? And further, how is it, given the vast amount of crucial intellectual work done in fields that critically investigate narrative forms and strategies, that it is still binary life stories—even those proven to affect the most atrocious chapters of human history—to which theorists such as Latour return? As yet another example of fatal attraction,

I argue that this is reason enough to ask what it means for literary scholars to do theory, with regard to *life writing* in particular. How do scientific and legal devices cut and tailor *life* into its forms, and what is the corresponding narrative logic that constantly legitimizes these practices? How does this logic uphold the life story, what forms and genres can it take, and what does it mean to be "represented" in these all-too-real narratives? What has (literary) theory done, and what can it do?

12

# The Right to Narrate (a) Life

CUTTING THEORIES

Life is not your history. —GILLES DELEUZE AND CLAIRE PARNET, *Dialogues II*

That is a rather abstract way to narrate a story, this fable you jealously call your story, a story which would be solely yours. —JACQUES DERRIDA, *The Monolingualism of the Other*

All biographies like all autobiographies like all narratives tell one story in place of another. —HELENE CIXOUS, *Rootprints*

In mobilizing Haraway's language of the *life story*, I do not seek to point to a shared fictionality at the core of law and of literature but rather to expose the violence of this particular image of thought and the genres it creates to legitimize its conceptual and actual slicing of life and lives. Literary scholars have often emphasized the complicity between genre and law. As Derrida writes in his "Law of Genre" essay, as soon as the word *genre* is articulated, heard, or written, law appears too, in the form of limits, interdictions, and orders.[1] Joseph Slaughter goes even further, suggesting that *law* "enables some narrative plots and literary genres over others," while *literature* "has historically favored and enabled some formulations of law."[2] Taking his argument seriously means addressing the complicity between law and life writing by disentangling our normative modes of storytelling. Indeed, the subject of traditional Western autobiography shares with the legal subject not only its Enlightenment-informed attributes but also the privileged position that allows it to claim a narratable and recognizable life, mostly to the detriment of other lives and modes of existence. It is no coincidence that literary scholars have rarely not touched upon law when writing about autobiography, while law and legal theory have in turn also relied upon generic notions of autobiography or anthropocentric biography when justifying how normative Cartesian cuts create clearly delineable and bounded lives.

While literary theory has witnessed a shift away from classical autobiography toward nonanthropocentric and posthumanist forms of life writing, law still feels most comfortable in its humanistic and anthropocentric framework. It is still the Cartesian notion of the thinking mind, Kant's transcendent subject, Anthropos—that is, the Enlightenment construction of the sentient, autonomous, rational subject, superior to body and nature—that operates at the very core of our conceptions of who or what can narrate, live, and have a right to something like *a* life. It is also still trust in representation over the physical world that forces bodies and matter into its fixed linguistic, legal, and conceptual frames. This in turn paves the way for analogy and comparison, which both construct a notion of difference that only confirms or rejects similarity to what has already been declared the superior category. Who or what has historically been able to fit the definition of legal subjectivity, and in which political and geographical contexts? Can a river, an insect, a monkey, an atmosphere, a digital mind have *a* life? Are these lives equal? And what is equality anyway in terms of (a) life? One way of attending to these questions, rather than presenting a solution, is to notice how legal and literary theory enforce each other, and how their alliance, rooted in the concept of the liberal subject, conserves an anthropocentric and neohumanist notion of law that by no means protects all human life, let alone nonhuman life or nonlife. I suggest examining the complicity between the theoretical approaches of two different yet inextricable disciplinary perspectives, attending first to how autobiography finds, even cuts, its way into law and consequently to how law is tied to the autobiographical genre.

## Autobiography in Law: The First Cut Is the Deepest

In her book on law and new materialism, *Law Unlimited*, Margaret Davies asks whether law can be understood beyond the subject-object distinction despite its conceptual and historical commitment to that framework and its typical perception as a solely human construct. "Can tangible stuff be anything other than an object of law's interpretive gaze?" she asks.[3] This is as much a question of *law* as it is of *storytelling*, for storytelling operates according to its own laws too. If the response to the story of the Anthropocene has brought about feminist, neomaterialist strategies that expose and rearticulate the entanglement of not only life-forms but also word and world, then the operation of storytelling can also be reimagined as cutting across material and immaterial, conceptual and real, and, in any case, entangled realms of matter and meaning production: "an electron crashes into a language, a black hole captures a genetic message, a crystallization produces a passion, the wasp and the orchid cross a letter."[4] The notion of a detached observer that classifies matter at will or of experimental arrangements that enact "a cut through which the 'instrument' and 'object' obtain well-defined boundaries and properties"[5] rests on the assumption of a clear separation and hierarchy of mind and matter. This is also why Haraway's claim that the life story must change in its *marrow* has to be read *matterphorically*. The question of *life*, even of our conceptions of it, can never be purely metaphorical or representational. Thus, speaking of the life *story* does not render it purely fictional, literary, or a construct of a human mind. Rather, it points to the material-discursive practices that constitute *life* by excluding other conceptions and definitions that nevertheless remain just as *real*. From this perspective, it is clear that theory *matters*.

Even more directly, however, Davies's question points to the anthropocentric perspective inhabiting Western law and legal theory, which is characterized by a specific temporality and spatiality. Within this perspective there is a world populated with mainly human or human-*like* constructs—subjects, persons, selves—that are sentient, conscious, rational, and display life interest, will, and self-autonomy. Here, analogy and comparison enable operations of inclusion or exclusion, as well as recognition or negation, that result in a corporation or a ship bearing enough similarity to a human to be considered legal persons, whereas a butterfly or a scooter does not. It is likeness as resemblance, imitation, and filiation that blocks other life-forms and existences from cohabiting in this constructed world. In order to create the desired population and reduce what is left to nonlife or property, law—as I will show—draws its lines, performs its cuts and surgeries, and enforces its limits onto concept and matter. While these operations are always potentially violent, it is crucial to note that the issue is not demarcation per se. The absence of demarcations would lead to homogenization—that is, the erasure of difference or differences. Homogenization is about reflection and imitation, being-the-same or being alike. As Barad reminds us, differentiating differently is a practice of cutting together-apart; it should be about not separating or othering but continuously making commitments and connections.[6] However, the right to (a) life—and the entire history of Western law—functions according to a reductive concept of difference and an overweighted emphasis on similarity: Each legal category, no matter how broad, attains its meaning solely by comparison, through opposition to another category from which all other categories are similarly excluded.[7] This inclusion or exclusion process operates as a sharp Cartesian cut and is indifferent to scale.[8] Even at the molecular level, it is performed by an "ontological surgery"—the carving up of nature "to alter collective perceptions of the meaning of life."[9] As Naffine states, it is still mainly "at the borders of human existence," the "edges of human life," that legal development, the limits of legal personhood, and the right to life are negotiated.[10] However, this does not mean that "human lives" are protected by a right to life; quite the opposite, as "those who die every day because of famine, sickness, and war" refute "the very pronouncement of a right to life."[11] This is why Esposito urges us to finally admit that "no right is less guaranteed today than the right to life."[12] Taking Esposito's observation seriously, we might wonder how cutting edge current legal theories that deal with *life*, even a right to (a) life, are, as well as what it means to be cutting edge at the intersection of law and life. The following three examples ((auto)biographical beings, biographical lives, and ontological surgeries) provide a glimpse at such intersecting and demonstrate that, when it comes to a right to (a) life, autobiographical laws—that is, laws that refer to life as the property of a self, that are narratable by this very self, and that separate life from nonliving, other clearly bounded lives, and objects—are not only conceptual applications. Rather, they are written into flesh; they are even engraved into molecular sequences or cut with molecular scissors.[13] It is here that we might wonder how encounters between theory and life actually produce cutting-edge concepts, including a right to (a) life.

*(Auto)biographical beings.* Warwick Fox's line of reasoning provides an excellent demonstration of how law cuts and territorializes life, and also how the genre of autobiography is complicit in these operations. Fox, best known for his work on ecophilosophy and deep ecology, acknowledges that humans are not the only sentient beings and argues that we should "seek to avoid inflicting

unnecessary pain and suffering on sentient beings in general and we should seek to avoid causing unwanted death to selves in particular."[14] In his widely quoted essay, Fox performs various conceptual cuts. The first separates nonlife from life as embodied by beings. The second sorts through sentient beings, separating nonhuman beings (only sentient) from human beings (selves). The third runs through (human) selves, separating "normal" and "healthy" selves from others by claiming that "normally developed humans are the only selves—the only beings with autobiographical self-awareness—that currently exist on earth."[15] Fox makes it seem almost natural and logical that a fourth cut follows from the preceding three. He claims that, although all kinds of sentient beings can be harmed by the infliction of suffering and pain, only selves—only *autobiographical beings*, as the quotation will show—can be harmed by the infliction of death per se:

> This is because only selves can, as it were, be cut off from themselves—from *their* own awareness of their existence; from *their* memory claims upon *their* past, *their* dreams, plans, and projects for the future, and their self-aware location of the present in that autobiographical context—and, thus, only selves can self-reflectively not want this to happen (or, in the case of, say, painful terminal illness, sometimes self-reflectively want this to happen). This means that unwanted death is a harm to autobiographical beings from *their* perspective and is mutually recognized as such by rational selves. In contrast, death *per se* does not cut sentient beings off from 'their' past, present, or future because they are not autobiographical selves; their death simply means that they die in this moment rather than that moment.[16]

The articulation of *life* as opposed to death, and the subsequent claim that life and death become a matter of *perspective* (autobiographical, self-reflexive, introspective) itself linked to a specific understanding of both serial temporality and its proper narration, states clearly that only a very selective group of sentient, healthy human beings can have *a* life, and consequently a right to (a) life. Indeed, only *autobiographical beings* are said to experience unwanted death since they are capable of being "cut off" from (an autobiographical) life. This line of reasoning not only denies "nonautobiographical" selves a certain kind of die-ability (death per se), it also relies on an optics based on reflection and anthropocentric mirroring. It is this logic, which underlies the classical image of thought, that Deleuze criticizes when he writes that representation connects "individuation to the form of the I and the matter of the self." The representation of the I, he continues, is not only "the superior form of individuation" but also "the principle of recognition and identification for all judgements of individuality bearing upon things."[17] For Fox, the autobiographical self is a thinking I, embedded in a temporal and spatial context that not only permits self-reflection and legitimizes a "perspective" but requires the recognition of other autobiographical selves. His argument reveals how the conception of life, in opposition to death, and its expressions as clearly delineable, self-owned property (*their* dreams, *their* consciousness, *their* memory, *their* past) mingles with the idea that the mutual recognition of this specific perspective is a prerequisite to possessing *a* life worthy of being legally recognized as such. Of course, the attribute "autobiographical" not only refers to a self-reflexive practice and perspective but has developed into a literary genre that has been crucial for normalizing the idea of a self that can clearly separate its life from that with which it becomes. This is also the reason why feminist, postcolonial, and post-humanist literary theory that critiques autobiography as the traditional genre of life writing has been influential far beyond the literary realm. Similar to Esposito's claim in regard to law and the right to (a) life, the literature scholars Sidonie Smith and Julia Watson emphasize the complex relationship between rights and personal life narratives by arguing "that the ownership of one's story has historically been less an intrinsic right than a site of contestation."[18] Indeed, the reductive notion of life enforced by the complicity between law and autobiography not only separates "human life" as a presumably stable category from other life-forms and modes of existence, but at times also cuts straight through it. Ultimately, no life, let alone what is considered nonlife, is safe from the potential violence of the cut that models lives according to readymade forms of the historically laden legal device of legal personhood/subjectivity or else reduces the remainder to objects—because it's not possible, Whitehead cautions, to "subdivide life, except in the abstract analysis of thought."[19]

*Biographical lives.* And yet, as shown by cases like that of the conjoined twins Mary and Jodie, born on August 8, 2000, this fact neither stops abstract thought nor restricts cuts that are enacted according to how that mode of thought imagines the world.[20] In this case, a separation operation was predicted to most likely lead to Mary's death (who was described as the "weaker" twin), whereas Jodie's chances to live a "normal" life were described by doctors as very high. Against the parents' consent and based on the doctors' assessment of Mary's and Jodie's chances to live a liveable—meaning "normal"—life, the separation surgery was performed on November 6, 2000. As expected, Mary died. The case is complex and controversial, and it is worth taking a closer look at the case documents, the legal reasoning, and the rhetoric used. In his ethical analysis of the judgment in the *Re A* case, John Harris, a transhumanist and bioethicist specializing in medical jurisprudence, supports the surgery on the basis that the twins lack legal personhood and, consequently, a right to life. Harris argues that what truly matters is "not life, nor yet human life," but rather a "certain cognitive capacity necessary to sustain a biographical life."[21] This perspective necessitates a pause for reflection: It is not simply life, nor human life, but the *narratability* of a specific notion of (more than) human life that holds significance. Because neither of the twins "had started living biographical lives," they cannot be considered persons. In this analysis, Mary did not lose a life.[22] Harris writes: "Persons, properly understood, are then individuals with a biographical life, individuals with full moral status in a way that non-persons are not. Persons, whom I believe are characterised by possessing the capacity to value existence, can be harmed by being killed or allowed to die because they thereby may lose something they value. Non-persons, which lack such a capacity cannot, by hypothesis, be deprived by death of something they could coherently be said to value."[23]

In this scenario, the surgical knives cut through not only life, separating subject from object, biographical being from flesh, but through what is understood as an embodied—perhaps all too embodied, too many bodies for one concept—human being. The "capacity to value existence" and the consequent ability to be harmed by death—i.e., to lose *a* life—are closely tied to naturalized, anthropocentric, and neohumanist assumptions about what a life, its temporality, and its alleged value are. In other words, the cut that guides the surgeons' is one that separates not only person from nonperson (literally even) but life, as self-conscious form of existence, from nonlife, as an existence that is presumably unaware of the "value" of its being. Here, the rights to have, claim, and narrate *a* life coincide uncannily, revealing their shared Cartesian logic and humanist ideas of selfhood and consciousness. This is especially interesting when seen in relation to Harris's fervid support of radical human enhancement and immortality (or longevity). For example, using Goethe's *Faust* to discuss the latter, he writes that there is "much evidence both from, and in literature, that many people are willing to trade off quality of life for longevity."[24] Trying to frame the desire to transcend death as both natural and rational, he states that "saving a life" means "postponing death," which is why extending life equals lifesaving. With reference to scientific developments and with a heroic undertone, he declares that "for the first time in human history we face the prospect of a truly open future, involving sequential as well as simultaneous opportunities," now "stretching, open-ended before the individual in an unprecedented but truly liberating pathway."[25] This echoes strongly with Max More's praise for the increasing possibility of "finally killing death, destroying the destroyer," and his articulation of the "deeply held values" of the extropian (transhumanist): an "enormously powerful will to life," a distain for death as the destruction of the self, and the belief that "those who cannot see why death is a 'bad' thing are sick."[26] This blends well with Harris's support for a moral obligation to pursue (human) enhancement, "new eugenics," and gene editing in human embryos; *life* in transhumanist and liberal understandings has an ideal form.[27] Ultimately, it is only this form that is considered narratable and thereby worthy of a past anchored in Enlightenment humanism and an endless future—preferably in outer space.[28]

*Molecular life writing: ontological surgeries.* However, it is not only in gene editing that the cut that cuts life into being and nonlife out of existence is performed on a molecular level. Precisely because decisions about what life is and when (a) life starts are made in relation to self-consciousness, the ability to feel pain, or the experience of "death," these questions are increasingly shifted to the molecular level—as if microscopic ontological cuts are less significant, ethically less challenging, as if normative cuts, which separate narrate-able lives from the storiless and rightless "tangible stuff," are more "reasonable" when manifested in microscopic ontological ones. Although shifting to the molecular might at first glance seem like an attempt to avoid morally charged discussions on the assumed value, beginning, or end of (a) *life*, the logic behind the cuts does not shrink or expand according to scale. Cartesianism, it turns out, does not dissemble under the microscope. On the molecular level, the decisions about where to draw the line between life and matter still result in both normative cuts and ontological ones. Indeed, law and life science remake the "facts of life" by means of ontological surgery. In fact, scientific laboratories are teeming with what Ingrid Metzler, a science and technology studies scholar, calls "strange bio-objects," meaning ordered parts of matter extracted from human and nonhuman life-forms.[29] Organized into modified plants and animals, immortal cell lines, or embryonic and nonembryonic stem cells, they await their legal status and classification. Legal tools such as patents on life-forms and copyright inscriptions into a synthetic microbe's genome rely on both ontological and normative cuts, reinforcing the law as a means of exploitation, extraction, and appropriation of both matter and meaning. On the molecular level, the cut that separates life from nonlife, which might be recognized as ontological, is still the very same that separates human from nonhuman life; the "auto-biography of man" is indeed carved into the *marrow* of the life story by Cartesian cuts. Autobiographical beings, biographical lives, and molecular structures whose potential for becoming part of this story is obscured all point to the fact that in the dominant, binary life story, *life* comes after the cut.

## Autobiography of Law: The Legal Subject
## That Therefore I Am

In law and legal theory, there is less attention to onto-epistemological cuts. As mentioned in the previous section, law is generally imagined to be an immaterial, abstract, humanmade system consisting of norms and rules articulated in language. The right to (a) life is mainly negotiated by means of a specific legal device: the legal subject or person. Through the concept of the person, Naffine attests, law helps define who *matters*.[30] While it does offer some creative potential within a certain realm, this device is highly divisive and prone to abuse. It can be mobilized to "enrich or to impoverish legal lives"; it can allow lives and life-forms to come into being, but it can also "unperson" altogether; that is, it can render certain (human and nonhuman) lives property and/or killable.[31] This concept's flexibility is tied to the underlying anthropocentric and neohumanist assumptions about what life is, which causes it serves, and which of its forms should have a right to (a) life. As has been argued extensively, the concepts of the person/subject/self in Western law are created in the image of the Enlightenment subject: endowed with specific properties, seen as an independent individual distinct from all other individuals, and willing to enter into contractual relations.[32] Among other developments that gathered momentum in Enlightenment philosophy, the Cartesian mind/body dualism, the notion of sovereign minds fostered by Kant, the social contract, and state of nature theories have determined how to think of law's (limited) obligation toward *life*, a concept that is itself unstable and shifting. Naffine and Anna Grear have provided indispensable insights into how certain legal theories have conceptualized the ideal legal subject as a disembodied contractor or a corporation, illustrating how legal theory has either erased the body of the legal subject altogether or imagined it as healthy, bounded, white, and male. They also explore how theories of will, property, individualism, and anthropocentrism have turned the legal person into a powerful instrument of exclusion and domination.[33]

A common response to the restrictions of legal subjectivity has been, and continues to be, the suggestion to further expand the given legal framework. As is well known, historically this has led to the inclusion of, for example, children, women, foreigners, and slaves into the exclusive circle of legal subjectivity, with at least a formal right to (a) life. Cases arguing for expanding legal personhood to certain animals, such as *Cetacean Community v. Bush* and *Naruto v. Slater*,[34] discussions on fetal personhood in the US; and attempts to extend legal personhood to AI or robots in the EU are recent examples of such an endeavor.

Although these attempts marginally broaden the narrow circle of lives and life-forms eligible for a right to life, they are nevertheless based on a test of whether or not certain entities bear enough similarity, or can be compared to, what law has defined as a person or subject. These efforts are thus questions of belonging driven by tree logic, the Cartesian cut, and the attempt to integrate into the legal framework what has already been declared a life-form or subcharacter in the binary life story.[35] As such, they challenge neither anthropocentric law and its imperialist and colonialist traditions and Eurocentric roots nor the normative scientific and conceptual life-writing practices of its authors. What's more, the deepest cut—namely, what brings the concept of life into being and separates it from nonlife—remains unquestioned. This means that not only are modes of existence perceived as devoid of vitality or as too different from human life rendered insignificant; *life* is also understood as an entity with clear boundaries that allows for operations of inclusion and exclusion. Although scientifically untenable, legal theory still refers to the philosophical idea of a clearly delineable (human) subject and its bounded body. Not only are dominant legal systems anthropocentric, generated as laws unto themselves in a glass homosphere; Cullinan points out that they are also based on the assumption that human beings only exist within their skin.[36] Animals and minerals, plants and animals, and photoautotrophs and chemoheterotrophs, Povinelli writes, can only be understood as external to the other if they are confined to a set of epidermal enclosures. As she shows, this imaginary separation can be debunked even from its very center. This is best demonstrated by asking where the human body is "if it is viewed from with the lung?"[37] What would be the genre, we might ask, that enables stories of life from the perspective of a lung? Or what concept of law would make stories of entangled existence narratable?

*Law of autobiography.* While it might seem to be a contemplation that primarily belongs to legal theory (informed by dominant biological, moral, and political notions of life), the forms of life writing and life narration have significant consequences for how life is conceptualized in law. The classical genre of autobiography consists of a particular form of life writing, specifically one that is authored by a subject that has a *right* to life, to *a* life, and to its authentic narration. The author makes a case *for* and a claim *to* a life. The fallacies of the self-sufficient, sovereign subject claiming to truthfully narrate an individual life have already been exposed by critical scholarship on autobiographical writing. As a result, literary theory and studies have argued for a shift from autobiography to other forms of life writing that offer alternative understandings of the self, the subject, and its relation to its environment and even propose discarding the concept entirely.[38] Besides postcolonial theory, the arguments that diverge most radically from the traditional subject evolve from neomaterialist and (critical) posthumanist life writing. Here, the aim is to understand and create the subject—as far as the concept of the subject can remain—not only as multiple but as materially embedded, embodied, and interdependent on other life-forms and modes of existence. In a posthumanist reading of autobiography or life writing, Stefan Herbrechter observes, "the (grammatical) subject or agent of the phrase can no longer clearly be disentangled from its object," which in turn "opens up the possibility for all kinds of post-anthropocentric forms of life writing to emerge."[39] When it comes to law and legal theory, however, the more pressure is exerted on its assumptions, the more the knot of law, life, and traditional autobiography seems to tighten. I argue that this is because, as a particular genre of life writing that is dependent on a particular notion of life (in contrast to nonlife), autobiography has developed alongside the very same concepts that uphold anthropocentric law too. As counterintuitive as it might seem that genre theory belongs here, the question is, according to Derrida, not one of belonging but of law. As stated above, as soon as the word *genre* is articulated, heard, written, the law—in the form of limits, interdictions, orders—appears too.[40] This complicity between law and genre repeats itself on multiple levels. In his broad study on the complicity between law and literature, and, more precisely, between human rights and the genre of *Bildungsroman* in nominating and mutually reinstating the "bourgeois white male citizen to universal subject," the literary scholar Joseph Slaughter demonstrates how law and genre are discursive regimes that create and test the subject of rights.[41] Taking this argument further, I would claim that, when seen in relation to how *life* is constructed, law and genre are not only discursive regimes; they also operate via material-discursive practices. If seen in relation to autobiographical writing—requiring a sovereign subject and a conscious self for whom *life* is a property—the law of genre and its narrative norms do not allow for nonlife to inhabit the position of an author, let alone to narrate *a* life from its perspective, without being spoken for.

It is therefore not surprising that, as Michael Ryan writes, autobiography is "the literary mode that became dominant at the same time (the late seventeenth century) as the liberal ideal of personhood."[42] The concept of legal subjectivity—a subject that has a right to its (self-owned) life and declares it as such—is at the core of autobiography. It "functions as an exclusionary genre against which the utterances of other subjects are measured and misread" and "provides the constraining template or the generic 'law' against which those subjects and their diverse forms of self-narrative are judged and found wanting."[43] An unforgiving genre, "auto-biography does not include degrees: it is all or nothing," Philippe Lejeune argues. Keenly aware and exceedingly appreciative of the proximity between autobiography and law, he famously claimed that "the autobiographical genre is a *contractual* genre."[44] Here, legal subjectivity is already invoked. If autobiography is a contract, it requires at least two legal subjects—the author and the reader—to establish the genre: a gesture that remains the anthropocentric and internally fractured prerogative of a certain legal subject, be it the author or the reader, for the tradition of Western autobiography serves to reproduce a particular subject or self; namely, that of universal Man.[45] Smith and Watson emphasize that, in traditional autobiography, "all 'I's are rational, agentive, unitary," which means that "'I' becomes 'Man,' putatively a marker of the universal human subject whose essence remains outside the vagaries of history."[46]

Although diverging with Lejeune on his theory of autobiography, Derrida also draws a line from autobiography to law. If autobiography were simply a genre, he claims, its single merit would be "that of permitting whomever speaks of himself to find refuge—in order to decline all responsibility and all onus of proof—behind the artificial authority of a genre, behind the right to a genre whose literary pedigree, as we well know, remains problematic."[47] For Derrida, too, pure autobiography does not know degrees, because "discharged of every onus of proof, pure autobiography authorizes either veracity or mendacity," and it does so "in accordance with a scene of witnessing, that is to say, an 'I am telling you the truth' without shame, bareback, naked and raw."[48] It is no coincidence that, in *The Animal That Therefore I Am*, Derrida calls into question the "auto-biography of man" as the auto-definition and auto-situation of Man that heavily depends on both "what is living and animal life."[49]

Nathan Straight makes a similar claim in regard to contemporary autobiography criticism and its meaning for what transcends the human, stating that it is "limited by its preoccupation with the autobiographical persona—the human self—at the expense of, or with peripheral attention to, a more-than-human world."[50] The pitfall of operating with the "anthropocentric, consumptive, and aggressively individualistic worldview" is, per Straight, that any contact with what lies beyond the human "must largely perpetuate antagonistic narratives of the self-in-contest with 'others'—both human and non-human."[51] If the right to autographical writing rests on the assumption of a human self that can narrate its story while deliberately ignoring that it depends on—and is in fact constituted by—what is considered the nonhuman, then autobiography is not even a contract, which would presuppose legal subjectivity on both ends, but an act of sovereign law making and life making.[52] As a consequence, this "auto-biography of man" does not permit the nonhuman (let alone nonlife) to be an autobiographical or legal subject, and it either requires its readers to agree to these terms or imposes them on the them. This becomes evident when Derrida, a few pages later, reflects on the Universal Declaration of Animal Rights, which, if considered even a step away from anthropocentric thinking, still falls short of questioning "the very idea of right, of the history and concept of rights, which, until now, in its very constitution, has presumed the subjection, without respect, of the animal."[53] Derrida points out that the history of the legal subject is inseparable from the history of the concept of the subject, who, while founding law and right, brought with them the denial of all rights to the animal.[54] Most importantly, this denial is not only restricted to the animal, as he states in an earlier interview with Jean-Luc Nancy entitled "Eating Well," published in *Who Comes After the Subject?* There, Derrida already relates subject to legal subject by pointing to the anthropocentric concept underlying both. He argues that the discourse on the subject continues to link subjectivity with "man" and does not grant subjectivity to the animal. Importantly, when asked by Nancy why he decided against limiting potential subjectivity to Man but then himself limits it to the animal, Derrida clarifies that his use of "animal" was solely for the sake of convenience and to use a classical reference; in fact, "nothing should be excluded."[55] Here, as well as in *The Animal That Therefore I Am*—the text that calls the "auto-biography of man" in question— Derrida reacts against a tradition of anthropocentric thought that restricts the indeterminate "who" to being assigned only to Man. In fact, Man stands in opposition to what is understood as the living in general, and Derrida notes that, "as long as these oppositions have not been deconstructed—and they are strong, subtle, at times mainly implicit—we will reconstitute under the name of subject . . . an *illegitimately* delimited identity, illegitimately, but often precisely under the authority of rights!—in the name of a particular kind of rights."[56]

This is a crucial observation as it does not simply open the concept of subjectivity for the nonhuman but, more importantly, goes further by indicating that the necessary shift for rights frameworks in legal theory is one of *kind* and not of *degree*. In other words, it does not suffice to create animal rights, environmental rights, or even human rights. The issue is precisely with the *kind* of rights, which are tied to a particular concept of the subject based on a "sacrificial structure" that consequently preserves the possibility "for a noncriminal putting to death" of whatever is seen as in opposition to Man.[57] Derrida shows how intricate the relation between autobiography—its roots, its concepts, its implications, and its cuts—and law is precisely because what is at stake is the impossibility of a right to (a) life that lies beyond Man's life story. Derrida importantly aims to deconstruct the cut that separates life from death, with its many political, social, ethical, and legal implications. Where his argument runs short, however, is that it does not address the fundamental separation between life and nonlife. He implicitly accepts the division between life and death to be the primary and ethically most egregious one when it comes to the construction of the legal subject and therefore ultimately also the right to (a) life.

Contemporary critiques of autobiography, informed by critical posthumanism and feminist neomaterialism, build on Derrida's foundational work by pushing it further to incorporate the concept of relationality as a material embeddedness within both *life* and existences. These critiques emphasize the interconnectedness of beings and the material conditions that shape their narratives, challenging traditional notions of autonomy and individuality. In *Material Self*, for example, Stacy Alaimo argues that "material memoirs" present a self that is "coextensive with the environment, trans-corporeal, and posthumanist" and demonstrates how the notion of selfhood is transformed by the acknowledgment that "the very substance of the self is interconnected with vast biological, economic, and industrial systems that can never be entirely mapped or understood."[58] Relying on Derrida's *The Animal That Therefore I Am*, Kari Weil makes us aware that we have, in fact, never been "fully or only human." Therefore, the task for posthumanist autobiography is "to take account of those who and that which have made us who we have been," finding forms that "give an account of those animal-others from whom I emerge," and in doing so, to develop a "hospitality of the self" that is aware of the many ways it affects and is affected by "that world and the many creatures of whatever name within it."[59] This corresponds with Cynthia Huff's claim that, while it is human life that has been central to biography and autobiography, it is precisely the destabilization of the "human centrality in favor of considering matter, the non-human, and the surround in which beings interact" that posthumanism offers in order to rethink the concept of autobiography.[60] If there is a notion of subjectivity that develops along these lines, it is probably closest to Braidotti's "critical posthuman subject," thought "within an eco-philosophy of multiple belongings" as a "relational subject constituted in and by multiplicity, that is to say a subject that works across differences and is also internally differentiated, but is still grounded and accountable."[61] Given the complicity between autobiography and law, can we expect law to shift its generic form of legal subjectivity in accordance with the shift of the self/subject in life writing? Would the "right to life" need to find a radically different legal expression, or would it perhaps expose itself as untenable without the framework of the liberal (legal and autobiographical) subject, with its right to (a) life? If the latter is the case, if life can be seen neither as an isolated entity, independent from all other modes of existence, nor a purely conceptual frame that allows certain entities to enter and forces others to remain excluded, then Western epistemologies and ontologies crumble—and with them, the accompanying notions of justice, subjectivity, human rights, democracy, and the crucial role that representation plays in all of them. Deleuze writes in *Difference and Repetition* that representation represents groundlessness as a "completely undifferentiated abyss, a universal lack of difference, an indifferent black nothingness." For representation, *"every individuality must be personal (I) and every singularity individual (Self)."*[62] If legal theory breaks with the idea of representation in classical (auto)biography, just as literary theory has challenged traditional modes of life writing—that is, of *life* and of *writing*—different stories will become possible. Of course, new-materialist and critical-posthumanist theories are just two of many possible ways to challenge dominant understandings of these concepts. In collaboration with Australian indigenous thinkers, Povinelli draws attention to the onto-epistemological cut between life and nonlife as articulated by the separation between biography and geography in Western epistemologies. In fact, she argues, "there is not biography (life descriptions) on the one side and geography (non-life descriptions) on the other," as they are "in a relation of extimacy": entangled and simultaneously internal and external to each other.[63]

Given the relation between life and theory that lies at the intersection of legal and literary theory, we are compelled to ask—perhaps in ways usually deemed unthinkable—whether or not *life* does actually protect (human and/or nonhuman) lives. Are these two uses of the term *life* comparable, or is it here that we find one of the most fundamental onto-epistemological conflations on which Western political and legal systems are built? To reiterate, it is certainly the case, Povinelli writes, "that the statement 'clearly, x humans are more important than y rocks' continues to be made, persuade, stop political discourse." And yet, it is "the slight hesitation, the pause, the intake of a breath that now can interrupt an immediate assent"[64] that demands close attention: a stretching toward, rather than a turning away. Any attempt to address this pause, the intake of a breath, must challenge representation, comparison, and the mode of analogizing thought that has used *life* not only as a makeshift bridge, as Ball recounts, but as a covering of the inexplicable and undeterminable.[65]

## Corporate Autobiographies

Whichever way we may choose to encounter these questions and concerns, law and legal thought follow a different logic altogether. Far from any attempting address posthumanist and new materialist thought, let alone thought rooted in non-Western epistemologies and ontologies, law and legal thought developed along the lines of, and remain dedicated to, neohumanist and transhumanist thought. Indeed, the posthuman subject in law is not to be confused with the posthumanist; the former preceded the announcement of the Anthropocene and cannot embody a life, let alone accommodate different manifestations of existences. It is the corporation and, by analogy, the ship, the self-driving car, the decentralized autonomous organization (DAO), AI, digital, sentient minds, all of which rely on the disembodied legal subject, that find instantiation in a corporate form rather than a human or nonhuman life form or form of existence. Anthropocentric and neohumanist law does not mean that only human beings can be legal subjects. However, in common-law theories, discussions over legal personhood are mainly found in the context of corporate legal theory, which again testifies to the complexity of the treatment of body, matter, and embodiment in law.[66] Grear emphasizes the paradoxical anthropomorphism at the core of corporate theory that, together with the disembodiment or hollowing out of the human being, produces the corporation as "the quintessential human subject" and the "ultimate instantiation of disembodied Anthropos."[67] As Richard Hardack observes, if corporations—corporate legal persons— were real people, "we would consider many of them pathological liars, and in some cases mass murdering serial killers: Exxon, Enron, Union Carbide, GE, Haliburton, Lockheed Martin, Monsanto, Philip Morris, Chevron, BP, PG&E, and, less directly, corporations such as McDonalds and Archer Daniels Midland, are responsible for taking and ruining tens of thousands of lives."[68]

Interestingly, Hardack also turns to the genre of autobiography to demonstrate how the autobiographical project of and in law is legible in the concept of legal personhood or subjectivity, which, as he demonstrates, ultimately consumes and cannibalizes life-forms and forms of life. He connects the concept of corporate personhood in US law to the specific form of autobiography it entails. Corporate speech, and especially advertising, creates what he calls "corporate autobiographies" (*corpographies*), which are "networks of representation that reify corporations as coherent and personalized entities, rather than treat them as legal fabrications."[69] Life writing, he argues "is an oxymoron in the corporate context—the corporation has no life, self, being, or agency, and its personhood exists only as a fiction and contrivance" that can be represented "only by partial surrogates, most obviously advertisers, officers, accountants," whereas "none of these figures are the corporation, or can author its autobiography."[70] And yet, "as these emblematic lives are exposed for public consumption, and effectively hollowed out, their subjectivities are symbolically transferred to the corporation, or contorted to satisfy the premises of corpography."[71] Echoing Paul de Man's description of the "autobiographical project" that produces the life it claims to narrate, Hardack reads advertisement as the manifestation of the corporate person that "fashions a self that is as corporatized as itself."[72] Closer to shadows or scarecrows than to life-forms, corporate autobiographies "do not describe or correspond to existing 'persons,' but generate them"; it is corpographies that actually write human lives.[73] On the one hand, "corporate personhood is a ghost in the economic and legal machine, but the ghost of something that was never alive"; on the other hand, it is "a contested site between corporations and people, with corporations finally, in effect, cannibalizing personhood."[74] A person for a person: equality via analogy and haunting formalism. What is crucial to note here is that it is not simply the corporation that displays forms of abstraction, cannibalization, objectification, and commodification. Rather, it is the concept of legal personhood or subjectivity in law that arises from the Cartesian mind/body separation—with its suppression of bodies and embodiments in a clear nonlife/life and object/subject cut—and the property relations that result from that separation, as well as its constant endeavor to create law in the image of Anthropos.[75] Not only has this particular concept in the past served to exclude, exploit, and kill millions and millions of human and nonhuman beings, let alone what was not even characterized as an existence or being, it also promises to continue doing so in the future. Bostrom—the British transhumanist and member of the *Extropy* email list that held the discussions that would shape contemporary transhumanism[76]—is widely celebrated for pointing out the possible dangers of AI. In his bestseller *Superintelligence: Paths, Dangers, Strategies*—abundantly praised by, among others, Bill Gates, MIT professor and transhumanist Max Tegmark, and Elon Musk—Bostrom imagines a future governed by AI and expresses his concerns about the treatment of sentient minds; that is, AIs or digital entities that possess consciousness and cognitive capacities comparable to human minds. He suggests that these digital minds could achieve a level of moral status due to their potential capacity to have conscious experiences and subjective awareness. Concerned about such digital workers, specifically about the "plight of working-class machines" situated in "a stable socioeconomic matrix that is already populated with other law-abiding superintelligent agents," Bostrom reveals the underlying equation of *life* with sentient (nonbiological) mind:

> Bringing a new biological human worker into the world takes anywhere between fifteen and thirty years, depending on how much expertise and experience is required. During this time the new person must be fed, housed, nurtured, and educated—at great expense. By contrast, spawning

a new copy of a digital worker is as easy as loading a new program into working memory. *Life* thus becomes cheap. A business could continuously adapt its workforce to fit demands by spawning new copies—and terminating copies that are no longer needed, to free up computer resources. This could lead to an extremely high death rate among digital workers. Many might *live* for only one *subjective day.*[77]

Further, Bostrom coins the term *mind crimes* and states that the erasure and destruction of simulated humans and (digital) sentient minds "might be equivalent to genocide." He even intensifies his concern by saying that "the number of victims might be orders of magnitude larger than in any genocide in history."[78] While technology continues to urge us to rethink laws and legal constructions, and while my argument is not that there should be no concern for digital workers, the implications of Bostrom's logic are consistent with the disembodied idea of legal subjectivity that aims to protect precisely the concept of law I wish to challenge. The violence in this flaw is exposed by the comparison he makes: What does it mean to state that the erasure of digital, sentient minds will be "larger" than any other genocide in history (meaning those that have actually taken place)? What serves as the ground for this comparison? Is the magnitude measured in numbers, and does it matter whether some numbers refer to (embodied) human beings and others to digital minds? And if not, how can Bostrom justify this equalization, given the history of atrocities committed by humans against humans and while, as Cullinan writes, exterminations of other species, living systems, and "even the life support systems of Earth" are to this day not even recognized as crimes?[79] Following a synaesethics of thought, what concept of (a) life is Bostrom projecting into the future by referring to "moral status," a normative concept that, as we have seen, operates as an ontological and epistemological knife, slicing the meaning of what (a) life is and can be into narrowly defined forms? How does this framework account for the deeply rooted ethical, historical, and ecological dimensions of such comparisons, especially when the erasure of embodied human lives and entire ecosystems continues to be dismissed in legal discourse?

As demonstrated in the first part of the book, transhumanists ultimately imagine a future that consists of disembodied minds, performing thought (or processing data) independently and detached from mortal bodies. Under this assumption, it seems reasonable, following Bostrom's logic, that a genocide of unprecedented magnitude might be possible without fleshly bodies. However, this neglects any acknowledgment of situated embodiments and their technological, material, and discursive relationality. The reductionist logic displayed here is in line with the extropian and transhumanist idea of an "ethics of egoism" built on the notion of the self as, to quote the transhumanist Bell, an information construct: "The self combines a particular set of memories with a particular set of thought processes, knowledge with intelligence, data with processing rules ... As I like to say, the self is an 'information construct.'"[80] This self—the extropian and transhuman self—is conceptualized as an idealized, postbiological iteration of the classical (legal) subject, wherein reasoning is redefined as high-speed computation and consciousness is understood as consisting purely of digitalized information.[81] Here, we might conclude, Descartes's *cogito, ergo sum*, along with all its underlying assumptions about reason, intellectual perception, and thinking, has reached its final stage.

From corpographies to transhumanist pondering on the rights of digital and artificial life, the autobiographical subject and the legal subject have developed alongside the concepts of the Cartesian subject and the enlightened self, gaining power from reason and, to refer to Irigaray, forgetful of aerial (and other) matter that sustains its existence. Perhaps, here again, it could be the pause, the intake of a breath, that indicates the rupture in a logic that has been too readily accepted. As paradoxical as it may seem, the quick equalization of a digital, sentient mind with an embodied human worker derives its legitimacy precisely from the possibility of an onto-epistemological cut, separating life from nonlife. This conceptual framework shares its roots with the unchallenged assertion in contemporary legal and political discourse that "x humans are more important than y rocks."[82] The cut does not guarantee that what is ultimately rendered as *a life* will be consistent with our current understanding of human life, nor with who we urgently aim to include in this inherently exclusive category. If the right to (a) life relies on an onto-epistemological separation between life and nonlife, then it is as much a right to exist as it is a license to efface existences.

# 13

# Matters of Indeterminacy

## STAKES OF CONCEPTING

Why is it that concepts and structures that are themselves indeterminate nonetheless still end up always on the side of the status quo? —MARTTI KOSKENNIEMI, *From Apology to Utopia*

If F = ma is formally empty, microscopically obscure, and maybe even morally suspect, what's the source of its undeniable power? —FRANK WILCZEK, "Whence the Force of F = ma?"

The alliance between law and autobiography—each of their foundations in Enlightenment concepts of property, sovereignty, reason, and consciousness—has been and continues to be a powerful conceptual liaison, sustaining a notion of the *subject* (as person or self) that asserts a *right* to *a life*. Grounded in transcendence and a concept of representation that insists on an inherent gap between word (or image) and world, this alliance determines the specific material-discursive practices that create and dissolve boundaries in relation to the virtual and actual fields of possible existences. Viewed through the lens of the concept of *life* it upholds, autobiography becomes a *contractual genre*. It contracts *life* into a single point, an entity that can be possessed and claimed, narratively delineated; it legitimizes the exclusion of that which does not bear enough similarity, let alone that which is incomparable. However, autobiography, as the classical genre of life writing, extends far beyond the literary or even the representational. To understand what is at stake, we can draw from the opening paragraph of de Man's widely cited essay "Autobiography as De-Facement." Here, de Man critiques the misconception that "autobiography is produced by life." Instead, he posits that "the autobiographical project may itself produce and determine the life," asserting that "whatever the writer does is, in fact, governed by the technical demands of self-portraiture and thus determined, in all its aspects, by the resources of his medium."[1]

The significance of understanding the implications of the generic, autobiographical traits in what has become a *binary life story* cannot be overstated. The reckless exhaustion of the medium's material resources (i.e., planet earth) and the technical demands of self-portraiture can ultimately render life forms as disposable property. Narrative superiority, strict (legal) monolinguism, property rights over life-forms, monopolies on scientific knowledge and tools, the tale of human progress, sovereignty, abstraction and extraction, formal equality, and generalization all dominate and record the contemporary anthropocentric and modernist legal atmosphere. The binary life story's narrative and prescriptive tools are Cartesian cuts, which veil the fact that every cut is ultimately an onto-epistemological one. From this perspective, Derrida's concern that we "never know, and never have known, how to cut up a subject" pertains to both language and matter.[2] It is a *matterphorical* concern because it falls in the middle, in the thickness of the cut, where word and world intra-act, and different stories, each incomparable to any other story, come into existence.

Law, however, continues to inscribe its power, carve out its subjects and objects, and narrate a binary life story through its claim of conceptual and linguistic transcendence. As Peter Goodrich notes in his *Advanced Introduction to Law and Literature*, "law seeks to portray a system of rules that are general in nature and abstract in their orderings." This self-image, he continues, "omits a key aspect of so-called black letter law, as also of jurisliterature": that "even its letters are material, coloured, haptic, housed and bound or now more often flickering across a screen but nonetheless spatially organized, material and mattering."[3] Philippopoulos-Mihalopoulos also cautions that law "does not dwell on the textual (that too) but expands on the space and bodies that incorporate it and act it out."[4] In other words, law is matterphorical, concepts matter-forth. To use Philippopoulos-Mihalopoulos's words, even though "law presents itself as immaterial, abstract, universal, non-geographical," it is nevertheless the case that "law is not just the text, the decision, even the courtroom."[5] It is "the pavement, the traffic light, the hood in the shopping mall, the veil in the school, the cell in Guantanamo, the seating arrangement at a meeting, the risotto at the restaurant."[6] The concept Philippopoulos-Mihalopoulos creates to describe this ontological continuation is the "lawscape," which is "mobilized by all its bodies and only by its bodies (nothing outside)," whereby some bodies have "a greater 'pulling' power than others."[7] The fiction according to which law is immaterial, textual, residing only in certain places, and applied only to certain areas, results, according to him, from a process of "in/visibilisation." In courts, as well as "prisons, concentration and refugee camps, nuclear heads, torture instruments, protests and revolutions," law is "fully visibilised and matter becomes a legal instrument," whereas invisibilization obscures the law in other realms, such as "bureaucracy, administration, obligations, ethics, morality, surveillance, health and safety."[8] In fact, law goes all the way down. "Transhumusian is its roots," it "lies beneath, slowly changing, the pre-nomos of the earth," mattering-forth, expressing synaesethically and jurisliterally the conditions of existence on this planet.[9]

This is why we must not excuse legal thought's commitment to analogy, metaphor, and comparison as an intellectual faux pas, innocent oversight, or unfortunate remnant of a time when thought could not imagine mingling with matter. Law, per design, categorizes, classifies, subsumes, includes, and excludes through its matterphorical concepts. It is with great precision that law continues to cut its subjects according to the requirements of its form, so that, as legal scholar Alain Pottage observes, "the juridical persona remains a prescripted or prefabricated role, and as such confirms the alienation of the very life or experience which it might be supposed to institute or enhance." What would it take to inhabit a law "without subscribing to the order which it silently reproduces, without adopting the masks or montages which law imposes?"[10] How can we reject the notion of the legal subject as this "anticipated conclusion offering a retrospective gift of meaning" and not just refuse to subdivide life according to pre-made forms of legal subjectivity but also question the cut that separates life from nonlife to begin with? Can *life* become graspable as entangled existence, and if so, how must legal concepts, including those of "rights" and "legal subjectivity," change in order to become response-able to this notion of (a) life?

In responding to these questions, we are tasked with practicing a mode of theory, especially legal theory, which, despite the inherent ambiguity (even indeterminacy) of *life* and the consequential cuts from which it results, remains ethically bound to the promise and expectations we read into the concept. We hold the expectations that justice, and its alignment with law, can manifest—at least at the moment when Eric Garner declares his inability to breathe, if not sooner—and that law is neither complicit in nor a facilitator of exploitative and extractive endeavors that ultimately destroy environments, atmospheres, and various forms of life and existence. We also maintain the promise that law and legal thought are not tools for enforcing Cartesian cuts but evolve from and are committed to a practice of immanent thought that fosters the "co-substantiality of forms of being" and acknowledges the "mutual involvement of all things in the immanent arrangement of existence."[11]

Yet it is not easy to be a traitor—not even to hesitate, pause, let alone prefer not to. How can we, and who among us can, approach the fundamental onto-epistemological cut without neglecting the necessity of having (a) life—not being in immediate danger of losing (a) life or being rendered nonlife—to even begin challenging it? Does pushing theory toward dismantling an image of thought so entwined with life and death, existence and nonexistence, endanger already vulnerable lives further, or does it finally make different kinds of existence possible? Perhaps it does neither, or even both. Either way, the status quo is unacceptable for some and, even worse, unlivable for others. This is not a question of ratio but of roots, rhizomes, entanglements that run deeper: response-abilities that do not transcend into morality but remain committed to an ethics of immanence. As a start, legal theory (and practice) must learn to think and to make sense-able different modes of existence without categorizing, classifying, or appropriating according to property regimes and without universalizing—that is, without flattening power and ignoring differential situatedness.

This brings us back to the conclusion of part II, emphasizing the significance of *matterphorics* as a mode of legal concepting and practice of theory. As I argue throughout this book, concepts, even when presented as fictional forms and representational vessels, must be approached matterphorically and in their historicity. Because concepts are "neither descriptions of existence nor transcendental forms of truth about existence" but "specific material doings or enactments of the world," they cannot be carried by the illusion—and violence—of fully discarding and replacing what has existed before.[12] Concepts come-*from* (historicity) and matter-*forth* (matterphoricallity).[13] *Terra nullius*, an international law concept, literally provided the ground for colonization and state sovereignty as the superior mode of community formation because its operation contained the existential negation of what it rendered unrecognized and nonexistent. Even if *terra nullius* were to return in relation to unclaimed digital territories, it cannot be understood detached from its history and its materializations: Where are all the electrons coming from, and what force is needed to use "bits to open innovation in atoms?"[14] The entanglement and historicity of legal concepts are also what Patricia J. Williams emphasizes in response to the critical legal studies (CLS) movement in the United States in the 1990s, stating that "in discarding rights altogether, one discards a symbol too deeply enmeshed in the psyche of the oppressed to lose without drama and resistance."[15] Wary of approaches that hold on to the possibility of a blank slate, a new beginning, or a replacement, matterphorics follows "the logic of the AND."[16] This is important not only in analyzing what existing concepts can do (i.e., their power) but also when seeking to create alternative concepts. Indeed, acknowledging already existing modes of matterphorical concepting, refusing to replace or ignore the workings of existing concepts (e.g., rights, legal person), is crucial for creating concepts that can operate differently. Important work, which performs embodied and situated strategies of subversion and creative complicity, is being done daily to materialize existing legal concepts differently. For more than a decade, indigenous activists and legal scholars have purposely mobilized the concept of the legal person/subject—a concept forcibly imposed onto indigenous understandings of relationality—by articulating "a new alliance between human and nonhuman forms of existence" and extending its field of workability to ecological entities such as rivers or nature.[17] By recoding relations of resemblance and facilitating inclusions (assimilations) of what has previously been rendered too different by legal thought, the question of "what happens to a region of existence when it is made a person within a liberal legal framework," as Elizabeth Povinelli puts it, actively rearticulates the legal concept's fields of workability. While the concept is not disregarded and still has to come to terms with its representational roots, such work is at least "slowing down the apparatus of extractivist capitalism."[18] It demonstrates the fact that concepts are "neither simply linguistic nor representational, but doings and becomings": matterphorical expressions that matter-*forth* and change the "dynamics of work-ability and non-workability in a region."[19] Yet it also shows the limits of how far concepts, still rooted in representationalism and an oppositional notion of difference, can be stretched. To work with the legal subject/person as a concept is less a choice than a necessity that stems from the formal incompatibility of conflicting normative orders and legal systems. As the concept determines who and what can be represented in, and therefore recognized by, the dominant legal system through its preexisting forms of signification (and valuation), complete noncompliance with and disregard of the concept means living in unrepresent-ability and unrecognizability—a state of existential impossibility in modern legal systems. For existence to be rightful in modernist legal systems, it must be mediated. The presumable choices offered are thus predetermined and render difference conflictual: either the exclusion, even negation, of indigenous modes of existence or creative complicity.[20]

This is also why Williams cautions against critical legal studies' hasty disregard of rights while also emphasizing the importance of not mistaking rights for a representational concept. Rights, too, must be understood matterphorically and in their historicity, constituted not by words but by their specific field of workability: by the lethal struggle of Black Americans in the United States. She argues that rights were given life by Black communities where there was none before: "we held onto them, put the hope of them into our wombs, mother them and not the notion of them." This was not, she emphasizes, "the dry process of reification, from which life is drained and reality fades as the cement of conceptual determinism hardens round—but its opposite."[21] It is precisely the question of righteous existence that still delicately hovers between its negations: between slave labor on sugar plantations and the bag of skittles the seventeen-year-old Black American Trayvon Martin bought at 7-Eleven shortly before he was shot by neighborhood watch volunteer George Zimmerman; between combat breathing on tobacco farms, the disproportionate impact of asthma and air pollution on the US Black, Hispanic, and indigenous populations, and the outcries of #icantbreathe following the killing of Eric Garner, who was stopped for allegedly selling untaxed cigarettes and put into an illegal chokehold by police officer Daniel Pantaleo until he suffocated.[22] Rather than disregarding the concept of rights, Williams suggests breaking its formalist, exclusive, and proprietary workings, stating that, "society must give them away. Unlock them from reification by giving them to slaves. Give them to trees. Give them to cows. Give them to history. Give them to rivers and rocks."[23]

Such distribution does not leave the concept untouched but rather shifts the power from the center to what the concept is capable of doing: "Flood them with the animating spirit that rights mythology fires in this country's most oppressed psyches, and wash away the shrouds of inanimate-object status, so that we may say not that we own gold but that a luminous golden spirit owns us."[24]

To "give away rights" does not mean granting or distributing rights, nor does it denote a gesture of generosity or belated responsibility toward the oppressed, invisible, rejected, or most vulnerable. The center must dissolve so that law can become immanent, entangled, and response-able to more than just the demands of imitation, filiation, and mimesis. The power in Williams's words lies in her prompt to give away rights like we give away secrets—secrets which have not been exclusively ours, which we can neither possess nor own—to break their spell instead of keeping them to us and for us only. Then, and only then, will they return to us, different, entangled, and more than just the sum of their reproductions of the center. To give away rights is to give away the right to autobiographical life writing, to draw a line between life and nonlife, and to split life even further, all too confidently, into subjects and objects, owner and possessed. What if rights are distributed, even decentralized, not through the distribution of the same form but through the enabling of decentralized (or a-centralized) participation in the concept's ongoing creation? What would it mean to write law differently and how would such writing—that is, creating, constituting—encounter *life*?

For Deleuze, writing, understood more broadly as creating, is not an expression of the subject and its centralized modes of sensemaking; it does not remain personal, does not produce reflections and representations, and cannot take (a) life as its object. Rather, according to Deleuze, writing is a "means to a more than personal life, instead of life being a poor secret for a writing which has no end other than itself."[25] He argues that writing cannot be personal but instead must "carry life to the state of a non-personal power." Only then does writing renounce "claim to any territory, any end which would reside in itself."[26] Rather than being "pure redundancy in the service of the powers be," and rather than remaining a matter of imitation, writing then becomes a matter of conjunction.[27] This explains why, for Deleuze, autobiography cannot have anything to do with life writing but rather strives to personalize, territorialize, and seize life. The author, he writes, "creates a world, but there is no world which awaits us to be created. Neither identification nor distance, neither proximity nor remoteness, for, in all these cases, one is led to speak for, in the place of . . . One must, on the contrary, speak with, write with. With the world, with a part of the world, with people."[28]

The writer "invents assemblages starting from assemblages which have invented him, he makes one multiplicity pass into another."[29] In other words, writing consists of more than the creative act of an author (or even a collective of authors) who produces a work, and it cannot be understood as a particular, isolated practice, separate from reading, speaking, swimming, or falling. What is important is the writer's relationality, the -*with* that creates meaning through intra-actions, to use Barad's term, and as becomings, to refer to Deleuze and Guattari. Thinking-with this concept of the writer, can we imagine the author of laws, the legislator, accordingly? If so, what would a *synaesethics* of legal concepting look like: one that not only decentralizes concepts but also fosters participatory normativity? How would it manifest on the ground, in atmospheres, falling and rising, mindful of air, and embracing indeterminacy?

# Epilogue: A Concept Yet to Come

## A DECENTRALIZED RIGHT TO BREATHE

It was precisely the questions at the end of part III, as well as my inquiry into how matterphorical concepting could shift the conditions of existence on the ground, that occupied my mind (and body) in 2020. In pursuit of an answer, or more precisely the construction of a concept, I cofounded LoPh+ and initiated its "Gas Exchanges and the Right to Breathe" investigation, which evolved into the "Decentralized Right to Breathe" (De.RtB) initiative. This ongoing, multiyear collaborative endeavor involves dozens of researchers, scientists, Web3 coders, AI architects, artists, lawyers, curators, scholars across various disciplines, social entrepreneurs, hardware and software engineers, friends, and more-than-friends. The timing of these inquiries, for the endeavor to collaboratively invent a novel type of matterphorical concept, was anything but coincidental. At that juncture, the ambiguity surrounding the concept of (a) life, the differential judgment on whose lives matter (and whose do not), and the relationship between legal concepts and differentially valued existences was

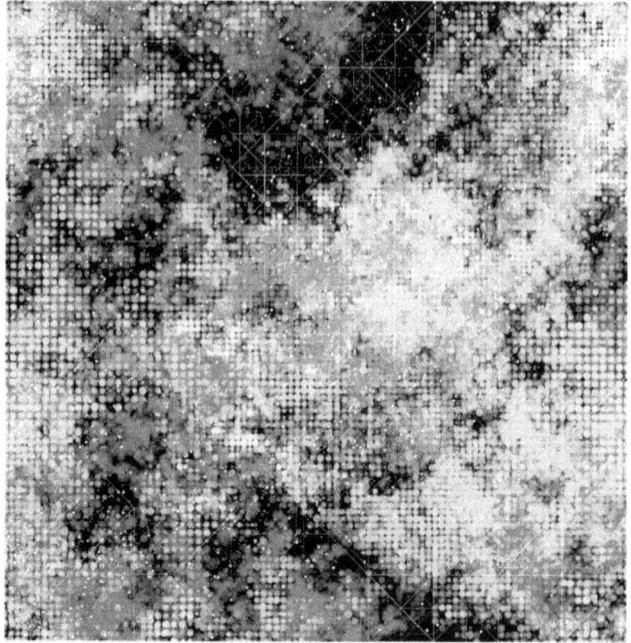

FIGURE E.I.
Visualization
of the author's
breath, minted
as an NFT in
OpenSea.

acutely pronounced. Issues of racial and economic injustice, governance failure, police violence, and global warming—all significantly altering modes of existence across scales—were articulated through unbreathable atmospheres, oxygen shortages, and respiratory arrest. Indeed, the outbreak of the SARS-CoV-2 virus, along with increasing concern regarding air pollution, the suffocation and murder of George Floyd by US police officer Derek Chauvin in Minneapolis, the reverberation of #icantbreathe across the screens and streets of the United States and beyond, and a cascade of natural catastrophes such as wildfires, hurricanes, and extreme weather events caused by human-induced climate change were palpably shifting the physical and social atmosphere of this planet. Even in regions where breaths were taken for granted, in their iterability and their sustainability, the question of what it means to breathe was raised in the most precarious manners, at times at the very threshold of non/existence. Breaths were held, face masks divided countries, ventilators were scarce. As the pandemic contracted more and more breaths, calls for legal protections for the act and conditions of breathing became louder and continually resonant. Efforts to curb breathing injustices invoked human rights protections in the form of a right to breathe clean air. Given that, according to the World Health Organization, 99 percent of the world's population breathes air that threatens their health, such frameworks are undoubtedly indispensable.[1] Attaching an object to the verb "breathing" (specifically, "clean air") is a strategic and practical decision to resolve, at least for the context of that legal form, the indeterminacy of breathing and breaths. Given these frameworks, however, synaesethical perspectives raise important questions, pushing traditional rights concepts to their limits. Staying not only with the trouble but with indeterminacy, how could the rigid and inflexible legal form of rights attend to a phenomenon as intimately entangled as breathing—and how would it not perform its cuts, separating the breaths that matter from those that do not matter? Can a universal right address breathing injustice(s), incomparable and yet interconnected, in their singularity? Is breathability restricted to air pollution, and would a human right not necessarily be confined to those included in the category of the human? Concerns about definition, demarcation, and applicability are, of course, not unique to a potential right to breathe clean air. Yet the particular matterphorical configuration of breaths and the inherent interconnectedness of natural, social, economic, and legal processes of breathing raise a different question, focused more on potential (*potentia*: the force of multiplicities, the power *to*) than power (*potestas*: the power of an authority; the power *over* something).[2] Could a breath, with its unclear boundaries, its cross-scalar constitution from the molecular movements of oxygen molecules to atmospheric dynamics, uphold the conditions of *life* on earth without homogenizing it? Could its subversion of the separation between interiority and exteriority perhaps guide the matterphorical expression of a different kind of right—nonrepresentational, collaborative, distributed, participatory, bottom-up, decentralized, and multiple? What would such concept entail? How would it matter-forth?

This book does not provide definitive answers to these questions. Instead, it closes by acknowledging that it has only reached a beginning: of thought that thinks-with, of *matterphorics* on and with the ground; of *synaesethics* as an ethics of sensing-with (what also) *matters*, including aerial matters; of a matterphorical case study on a decentralized right to breathe, materializing not as the story of this author but as written by a crowd as legislator. In fact, through its matterphorical case studies, concepts, acts of treason, ironies, and cautions, this book serves as a guide to what exceeds it and what it engenders. It is a beginner's guide not only to a book on law and its states of matter but also to a concept in the becoming (De.RtB), a preface for matterphorical concept creation in times of indeterminacy and an epilogue calling for creatively gaming complicity in what is increasingly referred to as the new governance frontier. In light of this indeterminacy, allow me to end with a glimpse into the former and indicate how *Matterphorics* enables a radical reconceptualizing of a legal concept. More concretely, I wish to show how it serves as a helpful guide in breaking the image of thought—the assumptions of what it means to think and how meaning comes into *being*—by investigating the notion of both "breathing" and "rights" in their matterphoricity.

Moving away from a proprietary mode of thinking, one that assumes that everyone already knows what it means to think and to breathe without considering aerial matter, the first question focuses on what breathing means. On a planet teeming with diverse forms of human, animal, and plant respiration, where the histories of gas exchanges surpass the capacities of history books and databases and where atmospheric shifts are inextricably linked to economic, social, and political processes, the question of what it means to breathe is far from self-evident. Where and when does a breath begin—can or should a right resolve this indeterminacy by forcing it into definitive forms? The temporal and spatial entanglements of breaths are vast and intricate; every second breath on earth originates from oxygen produced by marine phytoplankton, cyanobacteria, and algae during photosynthesis, the process which initially oxygenated the earth's atmosphere. As Raviv Ganchrow explains in the online article accompanying "Chlorophyll-Ocean-Iron-Breath," an art installation exhibited at LoPh+'s "Gas Exchanges and the Right to Breathe" exhibition, chlorophyll-bearing organisms contribute approximately 50% of the world's oxygen, play a crucial role in the carbon cycle, and form the base of the marine food web. Ironically, fossil fuels, currently diminishing breathable air, are remnants of ancient marine algae. Around 2.4 billion years ago, before the existence of oxygen-metabolizing lungs, dissolved oceanic iron combined with photosynthesized oxygen created global seabed deposits of iron oxide in what is known as the Great Oxidation Event (GOE).[3] Breathing, with all its intricate interdependencies, transcends mere human existence, encapsulating a broader spectrum of life. Air, a composition of nitrogen, oxygen, argon, carbon dioxide, neon, helium, methane, krypton, and hydrogen, is but one medium through which this essential process occurs. Lobsters and brittle stars breathe differently underwater. Plants engage in gas exchange through the stomata in their leaves, a sophisticated mechanism that encompasses both respiration and photosynthesis. Through these dual processes, they not only sustain their own energy and growth cycles but also contribute fundamentally to the atmospheric balance, exemplifying the profound interconnectedness of all *life-forms*. And even if narrowly conceived of as pertaining to human breath-takers, the act of breathing is profoundly influenced by the social and political climates breath-takers navigate: the harrowing realities faced by Black individuals in chokeholds, victims of domestic violence strangled, and refugees suffocating and drowning in their quest for safety. Moreover, indigenous communities suffering from industrial pollution, children in impoverished regions enduring hazardous air quality, and protesters around the world facing tear gas further underscore that the ability to breathe is not merely a biological function. Understanding what breathing means, and how the intricate relationality of meaning and matter manifests in different kinds of atmospheres, reveals that if a legal concept can help establish conditions of breathability for various modes of existence, it cannot be achieved through the letter of the law, traditional forms of rights and personhood, or authoritative issuance. It requires thought—in this case legal thought—to *think* what has been rendered nonsensical, sense-less, and insensible. This is the task of *synaesethics*.

Second, this book offers an *ethics* of sensing and sensemaking that refuses to reproduce the image of thought. It renders thinkable a different concept of right that does not restore the "'rights' of the individual" but aims to "de-individualize" by means of multiplication and diverse combinations.[4] As demonstrated throughout the chapters, modern law is based on representational modes of legal thought and thus rests on the construction of generality and equivalence through a syntax of resemblance, representation, and subsumable particularities. However, breathing in/justices cannot be subsumed by a generality of particularities—which is why the question of the applicability of a particular form, be it code or language, is one that negates the injustice in its specificity and singularity already through its premises. A right to breathe guided by a *synaesethics* cannot be a form, let alone operate according to a formula. It would have to relate not to representation but to repetition, which, per Deleuze, concerns "non-exchangeable and non-substitutable singularities":— echoes, reflections, twins, which "do not belong to the domain of resemblance and equivalence."[5] Such a distinction goes to the core of law—concerned with generality, formalism, and predictability—and the question of what a decentralized right to breathe, as a matterphorical legal concept, can *do*. After all, what if not breathing, even if reduced to the narrow and commonsensical idea of human respiration, reveals the power of repetition as an affirmative expression of difference as being (or ?-being)?[6] *A breath is not a breath is not a breath*. Each instance of breathing negotiates, yet affirmatively expresses, the existence of a singularity that is entangled with the social and material conditions of its possibility. Breathing is not an individual matter. No individual can breathe alone, detached from the social and physical atmosphere. As argued elsewhere, difference in breath is not expressed by a distinction between different breaths; it does not describe "the relation between pre-given entities and thereby [reinscribe] hierarchies and value-judgments."[7] Rather, "each difference passes through all the others" and "finds itself through all the others."[8] There is no consensus in breathing, only resonances. #icantbreathe is not simply a demand for legal change or a critique of a binary atmosphere spoken with one voice. It is the cacophony of multiplicities, the echo of breathing injustices, a sounding of "another justice, another movement, another space-time."[9]

Finally, *Matterphorics* is a book on power—on forces, attraction, gravitation, molecular desires, territorial sovereigns, intellectual love affairs, and mind fucking. It spans from the highest reaches to the lowest depths, tracing indeterminacies and inhabiting frontier spaces—not to determine from above but to mobilize energies to break the image of thought, the centralization of power, the exclusivity of decision-making, and the autobiographical approach to law-(and life-)making. In this sense, *Matterphorics* provides tools for experimenting with alternative modes of concept creation on the ground. Imagining a decentralized right to breathe, it enables an inquiry into a right that is distributed not through the allocation of the same form but through decentralized participation in the concept's ongoing creation. It asks: What if it is not the representable subject that holds a right to breathe but the breaths themselves that carry rights beyond Cartesian dualisms of mind/matter and subject/object? How can we collaboratively work toward a right to breathe that operates through the multiplication of difference rather than the distribution of sameness? How can a legal concept, moving in blocks, words, waves, or vibrations, break the arresting chains of signification? Can legal concept creation be democratized rather than remain the prerogative of centralized or market-backed power? It is not representational thought but *synaesethics*—an ethics of sensing and sensemaking that decentralizes concept creation by affirmatively relating to difference and attending response-ably to how matter and meaning become together-apart—that moves beyond the critique of centralizations of power.

These are, indeed, beginnings, indicating the potential of a concept manifested by a thousand plateaus, communities, breaths, blocks, and matters. If thought from the perspective of modern law, traditional market mechanisms, and representational thought, a concept of rights as indicated in the last few chapters may well be unthinkable. Yet this is precisely the point. And besides, who would have *thought* of an alchemy of rights? Far from simply being representational constructs or legal forms, immense alchemical fires—matterphorical encounters, mad physical particles crashing into language, semiotic fragments rubbing shoulders with chemical interactions—were required to give rights, not only thought but life.[10] Neither life nor rights were simply given, let alone created by a central authority (God, Leviathan, State, Market, Technology,

Reason). Why would we not approach the creation, distribution, and operation of a right to breathe in a decentralized manner, from the perspective of its power(s), a thousand materializing calls for a right to breathe? As argued, such a concept of rights cannot rely on the iterative applicability of the same (form, content) to difference, on the generality of the particular, but rather depends on the "universality of the singular," the entanglement of difference.[11] It must be capable of attending to injustices that are not only unrepresented but unrepresent-able and unrecognizable by current means of legal sensemaking: too multiple and promiscuous to be subsumed under even the most flexible general law. Indeed, such a concept of rights involves "inequality, incommensurability and dissymmetry" rather than "equality, commensurability and symmetry." It materializes (again and again, but differently) as that which "puts law into question" and works "against similar form and the equivalent content of law."[12] A right to breathe capable of attending to difference is creative, affirmative, even *inspiring* because it continuously invents relationalities and can hold breaths without appropriating or generalizing. It breaks away from a politics that seeks to "restore the 'rights' of the individual," as the individual is the "product of power." It instead aims to "de-individualize" by means of multiplication and diverse combinations.[13] Needless to say, the challenges are significant, and the undertaking is profound. Nevertheless, this is not a reason to refrain from pushing legal thought—through a thousand breathing injustices and a thousand plateaus of conceptual creation—until a decentralized and participatory right to breathe becomes thinkable, even sensical.

The last word, however, must be with (synaes)ethics. Is it possible for a thought, a concept, or an initiative to fully escape the attraction of frontier dynamics—for complicity to remain light enough to be negligible to their gravity? Is any attempt to subvert, hijack, or rechannel capitalist processes already doomed to fail, already too tainted by the act of thinking from within? Whether this is the case or not, no pureness, no innocence, and no unaffected position from which judgment could be made exists. Complicity, too, is a matter of degree and response-ability; to reiterate Massumi's observation, there is simply no outside of the capitalist process, at least as far as survival is concerned.[14] The task is to engage in *matterphorics*, a practice of affirmative critique that creatively gets "down and dirty in the field of play"[15] and does not "become enamored with power."[16] Such practice is closer to the creative yet complicit theft of the traitor than to the operations of the moral judge or objective arbitrator. What a concept can *do* and how it matters-*forth* are inseparable from where it comes *from*.[17] The "'great discoveries,' the great expeditions," Deleuze argues, "do not merely involve uncertainty as to what will be discovered, the conquest of the unknown, but the invention of a line of flight, and the power of treason: to be the only traitor, and traitor to all."[18] Power attracts, and thus attention must be paid to the adhesive molar structures of the state and its organs, as well as the lures of the "entrepreneurial spirit" haunting frontier spaces: a spirit—a breath or breathing, according to its etymology—that has historically sought to determine, appropriate, and commodify resources, including those indispensable for life-sustaining gas exchanges, such as human respiration, and the atmospheres in which past, current, and future modes of existence are negotiated. Yet the game is not over. Trust in each other in working and playing for a resurgent world can be built, across borders and disciplines, despite boundaries, notwithstanding dissent, and unthreatened by difference. The game, which of course has never been a game, is far from over: "One has to go all in. Fail successfully. That is an art. That is a project, and it is a life, a living law."[19]

# Notes

### INTRODUCTION

1. Viveiros de Castro, *Cannibal Metaphysics*, 39.

2. Viveiros de Castro, *Cannibal Metaphysics*, 45.

3. Viveiros de Castro, *Cannibal Metaphysics*, 40.

4. Deleuze and Guattari, *A Thousand Plateaus*, 3.

5. Deleuze and Parnet, *Dialogues II*, 51–52.

6. Deleuze and Parnet, *Dialogues II*, 52.

7. Deleuze and Parnet, *Dialogues II*, 44.

8. Deleuze and Parnet, *Dialogues II*, 44.

9. Deleuze, *Spinoza: Practical Philosophy*, 13 (emphasis added).

10. Negri, *Savage Anomaly*, xxiii. To be clear, this book was not written from a position in which my life was threatened or my body imprisoned, a fact that I wish to emphasize also because this is not the case for all of my colleagues at Princeton University. In 2016, Xiyue Wang, a graduate student at the history department, was, amidst his approved archival research, imprisoned in Iran and, after Trump's election as US president, sentenced to ten years in prison for espionage in 2017. Wang was released as part of a prisoner swap in December 2019. See Laura Secor, "Her Husband Was a Princeton Graduate Student. Then He Was Taken Prisoner in Iran," *New York Times*, July 10, 2018, https://www.nytimes.com/2018/07/10/magazine/american-civilian-hostages-in-iran.html; and Michael Crowley, "In Prisoner Swap, Iran Frees American Held Since 2016," *New York Times*, December 7, 2019, https://www.nytimes.com/2019/12/07/us/politics/iran-prisoner-swap-xiyue-wang.html.

11. *Onto-epistemological* is a term I am borrowing from Karen Barad. It refers to "onto-epistem-ology," which denotes "knowing as a material practice of engagement as part of the world in its differential becoming." Barad posits this concept as an alternative to the ontology/epistemology binary. See Barad, *Meeting the Universe Halfway*, 89–90.

12. Deleuze and Parnet, *Dialogues II*, 17.

13. The term *response-ability* (denoting the ability to respond) is borrowed from Donna Haraway and Karen Barad.

14. Viveiros de Castro, *Cannibal Metaphysics*, 40.

15. Frantz, *Black Skin, White Masks*, 1.

16. Deleuze and Guattari, *A Thousand Plateaus*, 25.

17. Deleuze and Guattari, *A Thousand Plateaus*, 25.

18. Deleuze, *Difference and Repetition*, 129.

19. Deleuze, *Difference and Repetition*, 130.

20. Deleuze and Guattari, *A Thousand Plateaus*, 353.

21. Barad, *Meeting the Universe Halfway*, 118. My account here is informed by Barad and their quantum-physical notion of ontological indeterminacy, which builds on Niels Bohr and argues against Werner Heisenberg's epistemological uncertainty principle, in favor of ontological indeterminacy.

22. Tsing, "Natural Resources," 5100.

23. Massumi, *99 Theses on the Revaluation of Value*, 69.

24. See Goodrich et al., *Law and Literature*.

## 1. THINKING-WITH MATTER(S)

1. Haraway, *Staying with the Trouble*, 42–43.

2. Haraway and Creager, "Compost-Ography."

3. Haraway and Creager, "Compost-Ography."

4. Massumi, *Postcapitalist Manifesto*, 68.

5. Massumi, *Principle of Unrest*, 91–92; Massumi, *Postcapitalist Manifesto*, 69.

6. Deleuze and Parnet, *Dialogues II*, 41.

7. Deleuze and Parnet, *Dialogues II*, 45.

8. Deleuze, "Sunflower Seed."

9. Haraway, *Staying with the Trouble*, 43, 58.

10. Arendt, *Eichmann*, 49.

11. Haraway, *Staying with the Trouble*, 36 (emphasis added).

12. Haraway, *Staying with the Trouble*, 47. For a more detailed analysis of the meaning of thoughtlessness regarding future imaginations of law, see Gandorfer, "Embodied Critique."

13. Butler, "Death Sentences," 49.

14. Schmitt, *Antworten in Nürnberg*, 60. In his written statement, Schmitt adds that the danger of theories, theses, and formulations being misconstrued is particularly pertinent in international law, constitutional law, and political theories, especially in times of latent or open civil war. The question of "ideological responsibility," and more generally the question of whether ideologies and doctrines should be justiciable at all, remains a difficult one, even more so, Schmitt states, in view of technology and propaganda. As is known, Schmitt was acquitted and released from prison.

15. Frisch and Kuhn, *Biedermann*, 38 (emphasis added). See also Povinelli et al., "Mattering-Forth," 318.

16. Haraway, *Staying with the Trouble*, 36.

17. See also Haraway, "Situated Knowledges." Haraway speaks of "a deeper surrender" to "immateriality, inconsequentiality, or, in Arendt's and also my idiom, thoughtlessness" in regard to what characterizes Eichmann's logic. Haraway, *Staying with the Trouble*, 36.

18. Waldron, "Thoughtfulness," 11.

19. Barad, "On Touching," 207.

20. Mbembe, "Decolonizing the University," 32.

21. O'Connor, "Space Colonization," 11. Max T. O'Connor decided to change his name to Max More to further emphasize the extropian ideal of progress and enhancement. For the sake of readability, I will refer to O'Connor as Max More, even in the cases where he has published under O'Connor.

22. O'Connor, "Space Colonization," 12.

23. Max More, "Transhumanism," 11.

24. O'Connor and Bell, "Introduction," 2.

25. Deleuze, *Foucault*, 119.

26. Harney and Moten, *The Undercommons*, 38.

27. Harney and Moten, *The Undercommons*, 31.

28. Goodrich, "Love of the Law," 344.

29. Foucault, "What Is Critique?"

30. Irigaray and Marder, *Through Vegetal Being*, 6–7.

31. See Gandorfer and Ayub, "Introduction: Matterphorical," 2.

32. Foucault, "What Is Critique?," 385.

## 2. METAPHOR AND NUCLEAR EQUATIONS

1. Haraway, *Staying with the Trouble*, 1.

2. Mawani, *Across Oceans*, 20.

3. Mawani, *Across Oceans*, 21.

4. Philippopoulos-Mihalopoulos, *Spatial Justice*, 6.

5. Philippopoulos-Mihalopoulos, *Spatial Justice*, 2.

6. Deleuze and Guattari, *A Thousand Plateaus*, 353.

7. Kniffin et al., "Beauty."

8. I am indebted here to Karen Barad and their insights they generously shared with me in the course of our close reading of this text, as well as to their talk at the *Reading Matters* conference in December 2018 in Princeton and their paper on material justice. See Barad and Gandorfer, "Reading Force(s)"; Barad, "End of the World."

9. Derrida, "No Apocalypse," 23.

10. Derrida, "No Apocalypse," 22.

11. Derrida, "No Apocalypse," 27. It should also be noted that Derrida's understanding of literature, in this text, is quite narrow—geographically and conceptually. In particular, Derrida has the dominant Western notion of literature in mind when he speaks of the "project of literature," which, he contends, arose contemporaneously to the principle of reason and "has not been possible without a project of stockpiling, of building up an objective archive over and above any traditional oral base." If Derrida fears the annihilation of the referent and the death of archive and literature, then this is what he has in mind. Derrida, "No Apocalypse," 26–27. See also Braddock, "Race," 95.

12. It should be mentioned that questions about the creation, function, characteristics, and potential violence of metaphor abound in philosophy and literary theory. Across the Western canon, and despite the differences in arguments, metaphor is mostly discussed

in relation to analogy, resemblance, and the transfer of meaning. As such, it constantly raises the question of the relation between what is considered real and representation, as well as that of the referent.

13. Aristotle, *Poetry*, 71–72.

14. Aristotle, *Poetry*, 73.

15. Aristotle, *Poetry*, 73. It is interesting to note that Derrida understands the "flames" as sunrays and, more generally, as light. However, his understanding of light is purely metaphorical and focuses on the shared topology with God, gold, value, knowledge, and Enlightenment rather than its physical characteristics.

16. Derrida, "White Mythology," 53.

17. Derrida, "White Mythology," 23.

18. Derrida, "White Mythology," 18; Derrida, "Force of Law," 14.

19. Derrida, "Force of Law," 7.

20. Nietzsche, "On Truth and Lie," 46–47.

21. Derrida, "White Mythology," 17.

22. Derrida, "No Apocalypse," 23.

23. Derrida, "No Apocalypse," 23.

24. Harney and Moten, *The Undercommons*, 34.

25. Deleuze and Foucault, "Intellectuals and Power," 207.

26. Deleuze and Foucault, "Intellectuals and Power," 207–8.

27. Deleuze and Foucault, "Intellectuals and Power," 208.

28. Hijiya, "Gita," 124n3.

29. Barad, "Posthumanist Performativity," 802.

30. Barad, "After the End of the World," 528.

31. Barad and Gandorfer, "Political Desirings," 21.

32. Deleuze, *The Logic of Sense*, 99.

33. Deleuze, *Difference and Repetition*, 30.

34. Husserl and Schuhmann, *Ideen*, 205.

35. Deleuze and Guattari, *A Thousand Plateaus*, 65.

36. Barad, "After the End of the World," 527.

37. Deleuze and Guattari, *A Thousand Plateaus*, 65.

38. Deleuze and Guattari, *Kafka*, 22.

39. Deleuze and Guattari, *A Thousand Plateaus*, 70.

40. An earlier version of the essay appeared as chapter 8 in Barad, *Meeting the Universe Halfway*.

41. Barad, "Invertebrate Visions," 227.

42. Barad, "Invertebrate Visions," 227.

43. For Haraway, diffraction, as an "invented category of semantics," mobilizes optical metaphors crucial to and common in Western philosophy and science. Diffraction is the production of difference patterns and as such "might be a more useful metaphor for the needed work than reflexivity." Diffraction patterns, Haraway writes "are about a heterogeneous history, not originals." For Barad, however, diffraction is "more than a metaphor," as it "attends to the relational nature of difference." Importantly, diffraction patterns, "as patterns of difference that make a difference," are not only semiotic but can be understood as "the fundamental constituents that make up the world." At times,

diffraction phenomena are an object of investigation, while at other times, Barad states, they will serve as an apparatus of investigation; never, however, will they be both at the same time. Diffraction, in other words, "not only brings the reality of entanglements to light, it is itself an entangled phenomenon." Barad's mode of doing theory can thus be called matterphorical. Haraway, *Modest_Witness*, 16, 34, 268; Haraway, *How Like a Leaf*, 100–101; Barad, *Meeting the Universe Halfway*, 72, 73.

44. Barad, "Invertebrate Visions," 228.

45. Barad, "Invertebrate Visions," 232.

## 3. MATTERPHORICS OF ?-HUMAN(ISM)

1. Braidotti, "Critical Posthumanities," 1–2.

2. Posthumanist accounts of ontological relationality vary, not least because of the insufficiency of human knowledge to once and for all demonstrate and define this relationality. Barad, for example, as a quantum physicist, looks at quantum entanglements and the agential cuts that articulate matter and meaning in and of the world. Braidotti uses the term *zoe*, understood as the "dynamic, selforganizing structure" of human, nonhuman, and prehuman life, a "transversal force that cuts across and reconnects previously segregated species, categories and domains." Barad, "Posthumanist Performativity," 808; Braidotti, "Four Theses"; Braidotti, *The Posthuman*, 60; Braidotti, *Transpositions*, 37.

3. This is also where posthumanism and (feminist) materialism join forces, so to speak. See, for example, Anna Tsing's or Stacy Alaimo's work.

4. Barad, *Meeting the Universe Halfway*, 136, 331–32.

5. We will see later, however, that posthumanism has been appropriated by other theoretical strands too. Transhumanists, for example, use the term *posthumanism* to describe the final stage of a development from human (as Anthropos) to posthuman, postbiological colonizer. This is, needless to say, fundamentally different from critical posthumanism as defined by, among others, Rosi Braidotti, Karen Barad, and Elizabeth Grosz.

6. Deleuze and Parnet, *Dialogues II*, 41.

7. Deleuze and Parnet, *Dialogues II*, 45.

8. Bignall and Rigney, "Indigeneity," 166.

9. Braidotti and Bignall, "Posthuman Systems," 2.

10. Wilson, "Indigenous Research Methodology," 176–77.

11. Needless to say, the fact that indigenous thought, which is itself multiple, is mentioned here rather than any other mode of thought that departs in its dynamism from the Western representational modes is inextricable from the fact that the latter have begun to allow certain theories on indigenous thought, under certain conditions, to enter their institutions. This also pertains to the questions of how to ethically think-with what is unknown, even unthinkable to our modes of thought and of how what happens to unthinkable modes of thought once they are forced to become recognizable.

12. Povinelli et al., "Mattering-Forth."

13. Barad and Gandorfer, "Political Desirings," 19.

14. Foucault, *The Order of Things*, 373.

15. Deleuze, "Nomadic Thought," 260.

16. Derrida, "Following Theory," 20.

17. Those questions are, needless to say, also contemporary ones. Only recently, for example, in the course of the Vienna elections for mayor, the right-wing Austrian People's Party candidate Gernot Blümel advocated for making the ability to speak German a prerequisite for access to social housing (a concept close to the idea of social democracy, especially in Austria). His aim was to exclude immigrants and refugees from any possibility of affordable housing and the very idea of social democracy. David Krutzler and Rosa Winkler-Hermaden, "ÖVP fordert Deutsch für Gemeindewohnung," *Der Standard*, September 6, 2020, https://www.derstandard.at/story/2000119848722/oevp-fordert -deutsch-fuer-gemeindewohnung.

18. Deleuze and Guattari, *Kafka*, 22.

19. Deleuze and Guattari, *Kafka*, 19.

20. Haraway, "Situated Knowledges," 580.

21. Haraway, *Companion Species Manifesto*, 5.

22. Glissant, *Poetics of Relation*, 19.

23. Glissant, *Poetics of Relation*, 27.

24. Braidotti presents a slightly different approach, which can be seen as lying somewhere between Haraway and Barad. By confirming and pushing Deleuze and Guattari's attempt to break the imperialism of language and signifier, to seek a materially intense expression, and by building on their conceptual persona, she emphasizes the importance of thinking-with figurations. Although coming from a slightly different perspective than Haraway and Barad, Braidotti, too, not only challenges anthropocentrism, European humanism, and representationalism but actively seeks to articulate modes of thought to undermine them. In doing so, she is adamant in emphasizing that figurations, although they are also literal expressions, are by no means metaphors or representations of static, universal claims but "markers of more concretely situated historical positions." They are first and foremost a particular "style of thought," neither metaphor nor embodied being. However, Braidotti also states that there are other styles and expressions of thought that are explicitly nonhuman. In "Animals, Anomalies, and Inorganic Others," she writes that shifts in theoretical approaches lead to the fact that finally "the metaphoric dimension of the human interaction with others is replaced by a literal approach based on the neovitalist immanence of life." In focusing on the animal, Braidotti further argues that it cannot be "metaphorized as other" but instead "needs to be taken on its own terms." Animals are thus neither functional parts in teleological taxonomies nor metaphors, for they "partake . . . in an ethology of forces and of speeding metamorphoses" and "express literal forms of immanence and becoming." Braidotti, "Animals," 528, 30.

25. Luckily, we also don't have to eject our arm or develop scientific and technological tools that enable us to blend visually into our environment. The question is also not whether or not the brittle star has a consciousness or whether we can somehow inhabit or reconstruct brittle star consciousness. In fact, thinking-with the brittle star is not about Enlightenment ideas. Besides, can we really claim to predominantly live in "bright and complex habitats"? Will Dunham, "No Eyes, No Problem: Marine Creature Expands Boundaries of Vision," *Japan Times*, January 3, 2020, https://www.japantimes.co.jp /news/2020/01/03/world/science-health-world/no-eyes-no-problem-marine-creature -expands-boundaries-vision.

26. Deleuze and Guattari, *A Thousand Plateaus*, 274.

27. In 1974, the philosopher Thomas Nagel approached a related quandary—as well as the question of objectivity—by thinking about the question that gave his essay its title: What is it like to *be* a bat? Nagel focuses on the bat because it has an experience of what it means to *be* in the world as a bat and because, he assumes, his readers might be more unwilling to agree that wasps or flounders have this experience too. What is more, it is the bat's very particular relation to meaning and knowledge production—namely, via echolocation (i.e., the reflection of soundwaves), which is fundamentally different from humans—that renders it ideal for Nagel to make his point. His answer, in a nutshell, is that we, as human beings, simply cannot know what it is like to *be* a bat because we cannot experience the world from the perspective of a bat. The question, he clarifies, does not refer to an epistemological problem because in order for us to even "form a conception of what it is like to be a bat," we would have to embody its particular point of view. However, the separation between self and other is not all too clear since, as he writes, "the distance between oneself and other persons and other species can fall anywhere on a continuum." While Nagel's essay is understood as a plea for subjectivism, it does raise interesting questions about the relation between what Whitehead calls "mentality" and an embodied perspective (or point of view). As Eduardo Viveiros de Castro notes, "a perspective is not a representation because representations are properties of mind, whereas a *point of view is in the body*." This is also reminiscent of Jakob von Uexküll's account of worldmaking as the differential *Umwelt*-making (environment-making) of living organisms and beings, in contrast to the false belief that there exists only one world "in which all living beings are encased," with one and only one space and time for all living beings. Nagel, "What Is It Like to Be a Bat?"; Viveiros de Castro, *Cannibal Metaphysics*, 72; Uexküll, *Worlds of Animals and Humans*, 54.

28. Haraway, *Modest_Witness*, 16.

29. Deleuze and Guattari, *A Thousand Plateaus*, 70.

30. Guattari, *Anti-Œdipus Papers*, 418.

31. Deleuze and Guattari, *A Thousand Plateaus*, 142. I do not mean to suggest that the doing away with modes of thinking grounded in representationalism, the moralistic image of thought (as Deleuze calls it), will lead to modes that are essentially good and less prone to violence but rather that in order to challenge the violence(s) made possible by representationalism we have to think differently about an ethics of thinking and living that cannot be delayed, postponed, mediated, or represented.

32. Deleuze and Guattari, *A Thousand Plateaus*, 40–41.

33. Deleuze and Guattari, *A Thousand Plateaus*, 40, 45.

34. Barad, *Meeting the Universe Halfway*, 90.

35. Barad, *Meeting the Universe Halfway*, 93. It is worth noting that this does not mean that there is no potential for violence or injustice in other modes of territorialization that do not share the traits of colonizing thought (which has a particular history of its own). Rather, this simply means that that particular mode of thinking cannot take hold, for it relies on transcendence, the creation of naturalized and hierarchically structured binaries. Barad is careful in emphasizing the need for an ethics that allows us to address and remain accountable for the cuts. See, for example, Barad, *Meeting the Universe Halfway*, 393.

36. Barad, *Meeting the Universe Halfway*, 139.

37. Barad, *Meeting the Universe Halfway*, 246.

38. Barad, *Meeting the Universe Halfway*, 376–77.

39. Barad, *Meeting the Universe Halfway*, 197.

40. Barad, *Meeting the Universe Halfway*, 198, 340. This also changes the conception of embodiment. While Haraway, in "Situated Knowledges," remained hesitant to engage with the question of the ontology of knowledge and understood embodiment as a particular and situated point of view, Barad makes those questions central to their agential realism. Barad writes, for example, that "embodiment is a matter not of being specifically situated in the world, but rather of being of the world in its dynamic specificity." Barad, *Meeting the Universe Halfway*, 377.

41. Deleuze and Guattari, *A Thousand Plateaus,* 512.

42. Barad, *Meeting the Universe Halfway*, 180–81, 83–84.

43. Barad, *Meeting the Universe Halfway*, 151, 70, 80–81.

44. Barad, *Meeting the Universe Halfway*, 141. The ontological inseparability, as well as the shift from uncertainty to indeterminacy, rests on an interpretation worked out in detail by Barad in chapter 7 of *Meeting the Universe Halfway.*

45. Barad, "Meeting the Universe Halfway," 149.

46. *Matter* here is meant in its double meaning. The fact that Barad's work attends closely to what *is* and aims to draw our attention to both ontological assumptions and possibilities for understanding ontology differently does not result from a privileging of one over the other but ought to be understood in relation to what is at stake: challenging the hegemony of representationalism. For an in-depth discussion of that relationality, see Barad and Gandorfer, "Political Desirings."

47. Povinelli, speaking from her perspective as a teacher, describes the issue pointedly: "There have been a number of conversations that I have had about forming concepts. I will tell you about one in particular, in which someone asked me to meet for coffee and talk about our work. When we were having coffee, they said, 'I just want to talk about a concept that I was thinking of working with.' And I said, 'Oh, well, what's the concept?' It was some word—I don't remember now which one it was. 'Well, where's your concept *from*'—and I was doing this twirling of my hands, signaling something like 'what is it *from*-doing?' Because the concept, for me, is precisely about this constant *from*-doing. The person responded with 'well I don't know; I just think it is such an evocative word.' I was surprised. I did not know how widespread this approach of substituting the concept with a word—the term, the name—was . . . It is crucial to attend to the history and specificity of its entanglements." Povinelli et al., "Mattering-Forth," 302.

48. Povinelli et al., "Mattering-Forth," 304.

49. Povinelli et al., "Mattering-Forth," 310.

50. Barad and Gandorfer, "Political Desirings," 29.

51. Barad and Gandorfer, "Political Desirings," 26.

52. Deleuze, *Difference and Repetition*, 192.

53. I want to thank Janette Lu and Layla Varkey, with whom I had in-depth conversations about a matterphorical understanding of "spaces of negligibility" in the course of their investigations into "Gas Exchanges and the Right to Breathe," conducted with the Logische Phantasie Lab in 2020–21. See https://lo-ph.agency.

54. These questions are precisely what the matterphorical case study on "Gas Exchanges and the Right to Breathe," conducted by the Logische Phantasie Lab, aims to address. If breathing cannot be understood universally, then how can we rethink a right to breathe that is sensitive to and can account for the many specific injustices that make instances of (human and nonhuman) breathing impossible? See https://lo-ph.agency/gas-exchanges.

55. Deleuze and Guattari, *A Thousand Plateaus*, 274; Barad, *Meeting the Universe Halfway*, 178.

56. Deleuze, "Hyppolite's *Logic and Existence*," 18.

57. Barad, "Meeting the Universe Halfway," 164; Barad, "Posthumanist Performativity," 812.

58. Barad and Gandorfer, "Political Desirings," 17.

59. Barad, *Meeting the Universe Halfway*, 141.

60. Guattari, *Anti-Œdipus Papers*, 246.

61. Deleuze and Guattari, *What Is Philosophy?*, 59. It is precisely the difficulty of expression and the reductionism of language that make Deleuze and Guattari caution in *What is Philosophy?* that the "plane of immanence is neither a concept nor the concept of all concepts." Concepts, here, are understood to be philosophical constructions (even if not in the traditional sense, as they state at the very beginning of the book already), which is why the equation of the plane of immanence with concepts would engender their becoming "universals and losing their singularity," while the plane "would also lose its openness." Deleuze and Guattari, *What Is Philosophy?*, 36.

62. Deleuze and Guattari, *A Thousand Plateaus*, 70.

63. Recently, for example, Elizabeth Grosz, whose careful interpretation departs from mine here, has written on the plane of immanence. See Grosz, *The Incorporeal*.

64. Deleuze, "Immanence: A Life," 31; Deleuze and Guattari, *What Is Philosophy?*, 38.

65. In *Difference and Repetition*, Deleuze elaborates on his understanding of univocity, especially in relation to his ontology of difference. He writes, for example: "With univocity, however, it is not the differences which are and must be: it is being which is Difference, in the sense that it is said of difference. Moreover, it is not we who are univocal in a Being which is not; it is we and our individuality which remains equivocal in and for a univocal Being." Deleuze, *Difference and Repetition*, 39.

66. Deleuze and Guattari, *A Thousand Plateaus*, 70.

67. Deleuze and Guattari, *A Thousand Plateaus*, 70.

68. I have written more extensively on this, as well as the question of expression in regard to legal theory, elsewhere. See, for example, Gandorfer, "Breathing Law."

69. Barad, *Measure of Nothingness*, 12.

70. Barad, *Measure of Nothingness*, 13.

71. Barad and Gandorfer, "Political Desirings," 31.

72. It is noteworthy to point out that both Deleuze and Barad, independently of each other, thought difference and expression with lightning. Deleuze, *Difference and Repetition*, 28; Barad and Gandorfer, "Political Desirings," 47.

73. Barad, *Measure of Nothingness*, 13.

74. Deleuze, *Difference and Repetition*, 57.

75. Barad and Gandorfer, "Political Desirings," 43.

76. Deleuze, *Difference and Repetition*, 57; Barad, *Meeting the Universe Halfway*, 234. For Deleuze, difference is thus neither contrarian nor negative. As ongoing becoming, difference is affirmative; it is being and nonbeing, or, as he suggests, "?-being," which denies dualisms and fixed entities and expresses existence as an ongoing, affirmative ontological questioning. This ongoing questioning functions for Barad as an "ontological opening for taking into account that the questioning is part of the world and the reworlding of the world," and as such, it is an "integral part of mattering otherwise." Deleuze, *Difference and Repetition*, 74; Barad and Gandorfer, "Political Desirings," 46.

77. Barad, "Invertebrate Visions," 236. According to Haraway, diffraction is the production of difference in the world, and diffraction patterns "are about a heterogeneous history, not originals," as they "record the history of interaction, interference, reinforcement, and difference." Barad adds a material dimension to diffraction, making it apt for a viable nonrepresentational methodology that can counter those of reflection. They explain that "diffraction gratings are instruments that produce patterns that mark differences in the relative characters (i.e., amplitude and phase) of individual waves as they combine," while reflecting apparatuses (mirrors, for example) produce images of objects placed at a distance from the mirror. They also point to the fact that "under certain circumstances matter (generally thought of as being made of particles) is found to produce a diffraction pattern," which is to say that "diffraction effects have been observed for electrons, neutrons, atoms, and other forms of matter." The ethical dimension of diffractive reading comes from the attempt to read, think, and understand differences that matter. "*Diffraction is a material practice for making a difference, for topologically reconfiguring connections.*" Haraway, *Modest_Witness*, 16; Haraway, *How Like a Leaf*, 100–102; Barad, *Meeting the Universe Halfway*, 81, 381.

78. Barad, *Meeting the Universe Halfway*, 381.

79. "Ontological thought experiments" is a reference to Barad's claim that all kinds of beings are "doing thought experiments with their very being." Barad, "On Touching," 207–8.

### 4. UNREASONABLE CANONS

1. See "Humanities Sequence FAQS," Princeton University, accessed July 11, 2025, https://humstudies.princeton.edu/humanities-sequence. There is certainly a lot to say about the specific canon and the rationale behind the selection of the fifty-eight books and texts, four of which are authored by women. However, the crucial discussions about, for example, canonicity and decolonialization, as well as the importance of engaging critically with canonical texts rather than abandoning them altogether, are beyond the realm of this chapter. The main focus here is the question of what thinking is and how it takes (a) place. It involves institutions and institutionalized knowledge production but is not limited to that perspective.

2. Whitehead, *Modes of Thought*, 171.

3. Whitehead, *Reason*, 4, 9. See also "Alfred North Whitehead (1861–1947)," by Gary L. Herstein, *Internet Encyclopedia of Philosophy*, accessed May 2019, https://www.iep.utm.edu/whitehead/.

4. Whitehead, *Reason*, 10–11; Manning, "Another Regard," 323.

5. Whitehead, *Reason*, 1.

6. Whitehead, *Reason*, 4.

7. Whitehead, *Reason*, 24.

8. Whitehead, *Reason*, 30. Actualities (actual entities) are, for Whitehead, the "final real things . . . of which the world is made up." Thus, there is not anything "more real" behind an actual entity. Although there are different graduations of importance and different functions among actual entities, the principle that actualities exemplify positions all on the same level. They are, he writes, final facts, drops of experience, complex and interdependent. In other words, an actual entity can be understood as the unity ascribed to a particular instance of coming into form (concrescence). The ontological principle at the core is that each actual entity arises from decisions (cuts) for it and that, by its very existence, each provides decisions for other actual entities that supersede it. Each actual entity adds a condition to the whole universe. As such, actualization can be compared to Barad's concepts of the agential cut and the cutting-together-apart. Whitehead, *Process and Reality*, 18, 212.

9. Keller, *Cloud of the Impossible*, 176.

10. Whitehead, *Reason*, 32.

11. Whitehead, *Modes of Thought*, 42.

12. Whitehead, *Reason*, 34.

13. Whitehead, *Reason*, 34.

14. Manning, "Another Regard," 322.

15. Manning, "Another Regard," 323.

16. Although recently Whitehead's work has been of great interest to scholars investigating its relation to quantum physics (an endeavor I consider highly interesting), I am not claiming here that Whitehead's and Barad's physics interpretations are the same.

17. Deleuze and Guattari, *What Is Philosophy?*, 72.

18. Deleuze, *Nietzsche and Philosophy*, 100–101.

19. Kuokkanen, *Reshaping the University*, 3, 17.

20. Kuokkanen, *Reshaping the University*, 13–14.

21. Kuokkanen, *Reshaping the University*, 5.

22. Whitehead, *Modes of Thought*, 63.

23. Coetzee, *The Lives of Animals*, 93.

24. Deleuze, *Difference and Repetition*, 131.

25. Deleuze, *Difference and Repetition*, 134.

26. Deleuze and Guattari, *A Thousand Plateaus*, 374.

27. Deleuze and Guattari, *A Thousand Plateaus*, 375.

28. Deleuze and Guattari, *A Thousand Plateaus*, 375.

29. Deleuze and Guattari, *What Is Philosophy?*, 88.

30. Deleuze and Guattari, *What Is Philosophy?*, 37.

31. Deleuze and Guattari, *A Thousand Plateaus*, 374.

32. Deleuze and Guattari, *A Thousand Plateaus*, 377.

33. Deleuze, *Difference and Repetition*, 147.

34. Deleuze, *Nietzsche and Philosophy*, 108.

35. Deleuze, *Nietzsche and Philosophy*, 108.

36. Barad, *Meeting the Universe Halfway*, 148.

37. Deleuze and Guattari, *What Is Philosophy?*, 37.

38. Deleuze and Guattari, *What Is Philosophy?*, 38.

39. Barad and Gandorfer, "Political Desirings," 44.

40. Deleuze and Guattari, *What Is Philosophy?*, 54–55.

41. Whitehead, *Modes of Thought*, 36.

42. Deleuze and Guattari, *What Is Philosophy?*, 212. As already mentioned, for Deleuze, thought is not a logical operation resulting from the unity of a thinking subject. Rather "something in the world forces us to think," which is, however, "an object not of recognition but of a fundamental encounter," that can only be sensed. That which forces us to think is *imperceptible*, insofar as perception relates to our (human) sense. Yet it is sensible because it is *in* and *of* the world. Deleuze, *Difference and Repetition*, 139. This is why Deleuze and Guattari describe thinking here as contemplation and, ultimately, as sensation.

43. Barad, "On Touching," 207.

44. Barad, "On Touching," 208.

45. Deleuze and Guattari, *What Is Philosophy?*, 42. Although it might easily be interpreted differently, I do not believe Deleuze and Guattari are suggesting an anthropocentric model of thought. For the claim that one does not think without what does not think already implies that there are two modes of thinking involved and that these modes are constitutive of each other. As we saw with Whitehead's stone thrown into the pond, thinking-with is not a unidirectional practice.

46. Deleuze and Guattari, *What Is Philosophy?*, 38.

47. A more recent and telling example pertaining to modes of sensemaking in relation to what is yet to be known can be observed in regard to COVID-19 and the many metaphors used by newspapers and other media, government and nongovernment officials, and science writers to convey various urgencies, affects, reasons for new directives and laws, etc. The war and wave metaphors have gained traction, especially at the beginning of the pandemic. For a discussion on theory and sensemaking in regard to COVID-19, see Helmreich et al., "Doing Theory." For a detailed analysis of the wave metaphor in this context, see Jones and Helmreich, "The Shape of Epidemics."

48. Deleuze and Guattari, *A Thousand Plateaus*, 2.

49. Deleuze and Guattari, *A Thousand Plateaus*, xiii.

50. Deleuze, *Difference and Repetition*, 136; Deleuze and Guattari, *What Is Philosophy?*, 85, 88.

51. Deleuze and Guattari, *What Is Philosophy?*, 72 (emphasis added).

52. Deleuze, *Difference and Repetition*, 139.

53. The English translation of this novel, published in 1979, is titled "No Place on Earth."

54. Deleuze and Guattari, *A Thousand Plateaus*, 377.

55. Unless indicated otherwise, all translations are mine. Where I have consulted yet adjusted the English translation by Jan van Heurck, I cite it along with the German edition. See Wolf, *Kein Ort. Nirgends*, 100, 104.

56. Wolf, *Kein Ort. Nirgends*, 94.

57. Wolf, *Kein Ort. Nirgends*, 101; Wolf, *No Place on Earth*, 109.

58. Deleuze, *Difference and Repetition*, 55.

59. Wolf, *Kein Ort. Nirgends*, 30.

60. Wolf, *Kein Ort. Nirgends*, 94–95, 102.

61. Wolf, *Kein Ort. Nirgends*, 102.

62. Wolf, *Kein Ort. Nirgends*, 101. Wolf, *No Place on Earth*, 109.

63. Butler, *Frames of War*, 38.

64. This is expressed by Frank B. Wilderson III in regard to a certain concept of Black thought when he states in an interview with Saidiya Hartman that "the slave occupies the position of the unthought." Hartman and Wilderson, "The Position of the Unthought," 185. Yet it is also not exclusive to Black thought—at least not insofar as thought is understood as multiple, which means that it can neither be universalized nor acquire an attribute by means of oppositionality or similarity to another mode of thought. That is to say, what breaks the image of thought is not singular but multiple, and therefore different kinds of counterthought cannot be universalized into and subsumed under the condition of unthinkability. Each mode of thought, then, is a multiplicity—each carries its characteristics, histories, precarities, and potentialities with it. Each is incomparable and singular yet neither internally nor externally homogeneous and thus neither a fixed entity nor a universalizable concept.

65. Glissant, *Poetics of Relation*, 6.

66. Glissant, *Poetics of Relation*, 196.

67. Glissant, *Poetics of Relation*, 8.

68. Glissant, *Poetics of Relation*, 20.

69. Glissant, *Poetics of Relation*, 196.

70. Deleuze and Parnet, *Dialogues II*, 55.

71. Glissant, *Poetics of Relation*, 156. That, throughout history, attempts to counter the established mode of thought, whether positioned within or assumed outside of the realms of accepted canons, have been mercilessly punished with means that reached from reprimand and exclusion to death sentences and killings cannot be denied. However, what Glissant aims to argue here is that the unknown, as well as that which departs from the known and from what is recognizable, bears, despite the violence it faces, the potential for different futures. This, of course, does not mean that the nonviolent violence required to break the image and open up possibilities to think differently is comparable to the unspeakable violence that was—and is—afflicted in the course of racism, discrimination, and other expressions of hate.

72. Mbembe, *Critique of Black Reason*, 11.

73. Glissant, *Poetics of Relation*, 155.

74. Glissant, *Poetics of Relation*, 9; Moten, "Blackness and Nothingness," 739, 742.

75. Povinelli et al., "Mattering-Forth," 312.

76. Deleuze, *Difference and Repetition*, 119.

77. Mbembe, *Critique of Black Reason*, 179.

78. The law of thought, as Deleuze attests, finds its expression in the Kantian critique that has everything—"a tribunal of justices of the peace, a registration room, a register"—except "the power of a new politics which would overturn the image of thought." Kant's critique could have had the potential for new thought but in the end only gives "civil rights to thought considered from the point of view of natural law," reinstating its truths and preconditions, incapable of real difference. Deleuze, *Difference and Repetition*, 136–37; Deleuze and Foucault, "Intellectuals and Power," 138.

79. Barad and Gandorfer, "Political Desirings," 40.

80. Deleuze, *Difference and Repetition*, 167.

81. Glissant, *Poetics of Relation*, 1.

## 5. MINDS FUCKING, MAKING LOVE

1. Deleuze and Guattari, *What Is Philosophy?*, 69.

2. Povinelli, *Geontologies*, 28.

3. "World" is of course not synonymous with "earth," but it also, especially from the perspective of a philosophy of immanence that rejects the separation between interiority and exteriority, encompasses the earth Deleuze is referencing. In general, Deleuze and Guattari's question is not meant to narrow the sphere in which thinking takes place and to which it should be seen in relation. Rather the question concerning the relationship to the earth makes visible the falsely assumed gap between mind and matter. The question can be reformulated as: "What is thought's relationship to the lived and immanent world?"

4. It is worth mentioning, however, that transhumanism, as a phenomenon and mode of thought, has different histories in different places. For the history of transhumanists, here immortalists, in Russia, see, for example, Anya Bernstein's *The Future of Immortality*. The US strand of transhumanism, which is the focus here, is decidedly libertarian and anarchocapitalist and as such differs also from the northern European strand.

5. Deleuze and Guattari, *What Is Philosophy?*, 45.

6. Williams, *Race and Rights*, 163.

7. O'Connor and Bell, "Introduction," 2.

8. O'Connor and Bell, "Introduction," 9.

9. A. "Arch-Anarchy," 11.

10. More, "The Philosophy of Transhumanism," 4.

11. More, "The Philosophy of Transhumanism," 6.

12. O'Connor and Bell, "Introduction," 1.

13. More, "The Extropian Principles," 17.

14. O'Connor and Bell, "Mindfucking," 10.

15. O'Connor, "Space Colonization," 12.

16. O'Connor and Bell, "Introduction," 13.

17. Dawkins, *The Selfish Gene*, 22, 46.

18. Dawkins, *The Selfish Gene*, 192.

19. Dawkins, *The Selfish Gene*, 192.

20. Frommherz, "Memetics of Transhumanist Imagery," 148.

21. Frommherz, "Memetics of Transhumanist Imagery," 150.

22. Frommherz, "Memetics of Transhumanist Imagery," 150.

23. Frommherz, "Memetics of Transhumanist Imagery," 148, 161.

24. O'Connor, "What's Wrong with Death?," 23.

25. O'Connor, "What's Wrong with Death?," 22.

26. O'Connor, "What's Wrong with Death?," 23.

27. O'Connor, "What's Wrong with Death?," 28.

28. More, "Philosophy of Transhumanism," 15.

29. Frommherz, "Memetics of Transhumanist Imagery," 159.

30. Moravec, *Mind Children*, 4.

31. Wolfe, *What Is Posthumanism?*, xv.

32. Nick Bostrom, "The Transhumanist FAQ, 5.

33. Hayles, *Posthuman*, xi. Hayles's influential and important book was written in response to Hans Moravec, one of the lead thinkers of transhumanist thought. Her use of the attribute "posthuman" would, from a contemporary point of view—which takes the significant differences between and histories of posthuman, transhuman, nonhuman, and more-than-human thought into consideration—better be understood as "transhuman."

34. Bostrom, "The Transhumanist FAQ," 4.

35. More, "Philosophy of Transhumanism," 4.

36. More, "A Letter to Mother Nature," 450.

37. Fuller and Lipińska, *The Proactionary Imperative*, 1. In this book, Fuller and Lipińska, indebted to More and the early transhumanists, first underline that eugenics—now promoted under the "slightly more politically correct rubric of 'human enhancement'—has been crucial to transhumanist thought from its very outset and then add their strong endorsement of eugenics as the "foundational science of human capital." The transhumanists' ambition "to produce people capable of embodying our full humanity" is, per the authors, "a task that requires both education and eugenics." This is reminiscent of an early *Extropy* issue published in 1991 where More claims the importance of genetical engineering, "eliminating deficiencies and maximizing the offspring's mental health, physical capacities, and emotional stability," and emphasizes that it must be ensured that "the design of a new transhuman child and the family form itself are left up to the individuals involved." In order to control population growth—not because resources will be scarce, but because most people "prefer a more spacious environment"—he suggests devising property rights in order to "make child-creator bear the full social cost of their activity," which "could begin by removing tax-subsidies to education and welfare subsidies that encourage large families." Fuller and Lipińska, *The Proactionary Imperative*, 3, 64, 98; More, "Order Without Orders," 29.

38. More, "Extropian Principles 2.5," 10; More, "Futurist Philosophy," 11.

39. More, "Futurist Philosophy," 11; A, "Arch-Anarchy," 13, 17.

40. O'Connor and Bell, "The Extropian Declaration," 51.

41. What is more, already in the first issue the fascination with new frontiers is expressed clearly. It is especially their offering of an "opportunity to make a fresh start as we like" and "a chance to experiment with new social orders, new religions, and new ways of living," which has been "hampered by existing governments which lay claim to every inch of the planet," that make them a crucial concept for extropians. More, "Philosophy of Transhumanism," 4, 6, 10.

42. Bell, "Extropia," 36.

43. Bell, "Extropia," 38.

44. Bell, "Extropia," 37.

45. Bell, "Extropia," 39.

46. Bell, "Extropia," 37.

47. O'Connor, "Space Colonization," 11.

48. Bell, "Extropia," 35.

49. Bell, "Extropia," 38.

50. Bell, "Extropia," 36, 38.

51. See also Gandorfer, "Breathing Law." "Elon Musk Answers Your Questions! SXSW 2018," posted March 11, 2018 by South by Southwest (SXSW), YouTube, https://www.youtube.com/watch?v=kzlUyrccbos.

52. Bell, "Extropia," 38.

53. O'Connor, "Space Colonization," 11.

54. Bell, "Extropia," 38.

55. See Thomas W. Bell's faculty web page: https://www.chapman.edu/our-faculty/thomas-bell (accessed May 2019).

56. Bell, *Your Next Government?*, 189.

57. Bell, *Your Next Government?*, 186.

58. Bell, *Your Next Government?*, 186.

59. Bell, *Your Next Government?*, 187.

60. Bell, *Your Next Government?*, 188.

61. National Lawyers Guild, "Report," 2. Note that although the National Congress of Honduras approved the Organic Law of Areas of Employment and Economic Development in 2013, the ZEDE has not yet come to full fruition.

62. The report states: "Many fundamental rights of Honduran citizens who live within the borders of ZEDEs are not protected under the new ZEDE law. These rights include: the right to Habeas Corpus or Amparo 20, Article 183; the inviolability of a right to life, 65; guarantees of human dignity and bodily integrity, 68; the guarantee against the extraction of forced labor, 69; freedom of expression, 72; protections for a free press, 73; freedom of religion, 77; guarantees of assembly and association, 78, 79, and 80; freedom of movement, 81; the right to a defense, to court access, and to counsel for indigents, 82 and 83; and freedom from non-legal detainment, 84 and 85." National Lawyers Guild, "Report," 8.

63. Alford-Jones, *Special Economic Zones*, 2, 6.

64. Mark Sullivan, "Zedes."

65. Mark Sullivan, "Zedes."

66. Startup Societies Foundation, accessed February 15, 2019, https://startupsocieties.com.

67. See the website of the Seasteading Institute, accessed May 2019, https://www.seasteading.org. The term *seasteading* is a combination of *sea* and *homesteading*, a legal principle according to which ownership can be acquired over a natural thing by using it or building something out of it.

68. Bell, "Extropia," 38. See "The Eight Great Moral Imperatives," Seasteading Institute, accessed May 2019, https://www.seasteading.org/videos/the-eight-great-moral-imperatives/.

69. Peter Thiel, "The Education of a Libertarian," *Cato Unbound*, April 13, 2009, https://www.cato-unbound.org/2009/04/13/peter-thiel/education-libertarian/ (emphasis added).

70. Geglia, "Honduras," 353.

71. "Announcement," Competitive Governance Institute, accessed April 6, 2022, http://ojs.instituteforcompgov.org/index.php/jsj/announcement.

72. Deleuze and Guattari, *What Is Philosophy?*, 41.

73. See "A Decentralized Right to Breathe," Logische Phantasie Lab (LoPh+), accessed July 18, 2024, https://www.lo-ph.agency/dertb.

74. Haraway, *Staying with the Trouble*, 4.

75. Haraway, *Staying with the Trouble*, 1, 4.

76. Barad, Meeting the Universe Halfway, 393.

77. Thaman, "Thinking," 15.

## 6. DRINKING THE ~~KOOL-AID~~ RED BULL

1. In literature, the *Ständeklausel* is a rule from classical drama that dictates that characters must adhere to their social class, often requiring that tragic heroes be of noble birth and comedic characters be of lower status.

2. Adrian Cho, "Gravitational Waves: Einstein's Ripples in Spacetime Spotted for the First Time," *Science*, February 11, 2016, https://www.science.org/content/article/gravitational-waves-einstein-s-ripples-spacetime-spotted-first-time.

3. Barad, "No Small Matter," G108–9.

4. Jhering, "Im Juristischen Begriffshimmel. Ein Phantasiebild," 244–334. Rudolf von Jhering's "Im Juristischen Begriffshimmel" satirizes the abstract realm of legal formalism by imagining a "heaven of legal concepts," where notions exist in lofty detachment from earthly life, critiquing jurisprudence's disconnection from social reality and purpose.

5. See also Maxwell, *Wax Impressions*, 4.

6. See also Seeley, *Honeybee Democracy*.

7. Maxwell, *Wax Impressions*, 3.

8. Barad, "No Small Matter," G103.

9. Scott, *Seeing Like a State*; Johns, "From Planning to Prototypes"; Foucault, "Preface," xiii; Massumi, *Postcapitalist Manifesto*, 69; Foucault, "Important to Think?," 475.

10. Deleuze and Guattari, *A Thousand Plateaus*, 40.

11. See Gandorfer, "Down and Dirty."

12. Manning, *Politics of Touch*, 70.

13. Braidotti, *Posthuman Knowledge*, 124.

14. Braidotti, *The Posthuman*, 29.

15. I use the neologism *in-elementary* rather than *subatomic* as an attempt to avoid the notion of scale embedded in Western modes of thinking (be it in the sciences, the humanities, or law). *In-elementary* does not denote a specific scale that is situated in relation and compared to other scales (micro, macro, etc.) but rather problematizes this mode of thought. The prefix *in-* refers to both the immanence of everything that is and its relation to what we understand (according to historical contexts and worldviews) as elementary (*in-elementary* as within the elementary) and, at the same time, the impossibility of arriving at a stable definition of a most elementary (*in-elementary* as not-elementary). The latter builds on Barad's work on quantum mechanics presented in *Meeting the Universe Halfway*, where they not only argue for a semantic indeterminacy (rather than uncertainty) but also emphasize that indeterminacy is both semantical and ontological.

16. Davenport, *The Space Barons*. Musk's youngest son's name, "X Æ A-12"—or, as Californian law, much to Musk's and his partner's annoyance, does not allow numbers in

names, "X Æ A-XII"—still carries parts of his father's fascination for Lockheed Martin's A-12 aircraft, built for the CIA in the 1960s.

17. Mike Wall, "1st Mars Colonists Should Be 'Prepared to Die,' Elon Musk Says," *Space .com*, 2017, https://www.space.com/34259-elon-musk-first-mars-colonists-prepared-die .html.

18. Claire Grodon, "Elon Musk Says 'Probably People Will Die' in Mars Mission," *Inverse*, June 2016, https://www.inverse.com/article/16857-elon-musk-says-probably-people -will-die-in-mars-mission.

19. Elon Musk, "Making Humans a Muliplanetary Species," posted September 27, 2016, by SpaceX, YouTube, https://www.youtube.com/watch?v=H7Uyfqi_TE8.

20. *Oxford English Dictionary*, "terraforming (n.)," accessed August 25, 2025, https:// doi.org/10.1093/OED/1913538300.

21. Unlike the polar ice on earth, which is frozen water ($H_2O$), the polar ice on Mars also consists of a thin layer of dry ice; that is, $CO_2$ in its solid form.

22. "Elon Musk will need more than 10,000 missiles to nuke Mars—Roscosmos," *Tass—Russian News Agency*, May 12, 2020, https://tass.com/science/1155417; Elon Musk (@elonmusk), "No Problem," reply to @jeff_foust, May 17, 2020, https://twitter.com /elonmusk/status/1262076013841805312.

23. Herron, "Deep Space Thinking," 613.

24. Jakosky and Edwards, "Terraforming Mars."

25. Bruce Jakosky and Christopher S. Edwards, "Can Mars Be Terraformed?," *Scientific American*, August 2018, https://blogs.scientificamerican.com/observations/can-mars-be -terraformed/.

26. *PBS Think Tank*, episode 1292, "Think Tank: Debating the National Standards," April 21, 2002, transcript, https://www.pbs.org/thinktank/transcript1292.html (emphasis added).

27. Meghan Henry et al., *The 2019 Annual Homelessness Assessment Report (AHAR) to Congress*, US Department of Housing and Urban Development, 2019, 12–13, https:// www.huduser.gov/portal/sites/default/files/pdf/2019-AHAR-Part-1.pdf.

28. "Joe Rogan Experience #1470–Elon Musk," posted May 7, 2020 by PowerfulJRE, YouTube, https://www.youtube.com/watch?v=RcYjXbSJBN8.

29. Investor Advocates for Social Justice, *Notice of Exempt Solicitation Pursuant to Rule 14a-103*, U.S. Securities and Exchange Commission, June 29, 2020, https://www.sec.gov /Archives/edgar/data/1318605/000121465920007479/d826200px14a6g.htm.

30. "Joe Rogan Experience #1470—Elon Musk."

31. "Joe Rogan Experience #1169—Elon Musk," posted September 2018 by Powerful-JRE, YouTube, https://www.youtube.com/watch?v=ycPr5-27vSI.

32. "Axios—Elon Musk Interview (Season 1 Episode 4 )," posted November 25, 2018, by Poopie Doopie, YouTube, https://www.youtube.com/watch?v=4qUA3nNWyCg.

33. "Axios—Elon Musk Interview (Season 1 Episode 4 )."

34. Max Chafkin. "Entrepreneur of the Year, 2007: Elon Musk," *Inc.*, December 2007, https://www.inc.com/magazine/20071201/entrepreneur-of-the-year-elon-musk.html; Neil Strauss, "Elon Musk: The Architect of Tomorrow, *Rolling Stone*, November 2017, https://www.rollingstone.com/culture/culture-features/elon-musk-the-architect-of -tomorrow-120850/.

35. See, for example, the work of Irus Braverman, Cormac Cullinan Margaret Davies, David Delaney, Anna Grear, Renisa Mawani, Andreas Philippopoulos-Mihalopoulos, Marie Petersmann, and Alain Pottage, as well as Sarah Keenan and Hyo Yoon Kang's Legal Materialities Network (https://legalmateriality.wordpress.com/author/ipissues/).

36. Elon Musk, "Making Humans a Muliplanetary Species," posted September 27, 2016, by SpaceX, YouTube, https://www.youtube.com/watch?v=H7Uyfqi_TE8; Barlow, "Declaration," 30.

## 7. *I AM* FREE, FREE FALLING

1. "Felix Baumgartner: First Person to Break Sound Barrier in Freefall," Guinness World Records, accessed May 13, 2019, https://www.guinnessworldrecords.com/records /hall-of-fame/felix-baumgartner-first-person-to-break-sound-barrier-in-freefall. The website also lists the following records: highest altitude untethered outside a vehicle, *first human to break the sound barrier in free fall, largest balloon with a human on board*, and most *concurrent views for a live event on YouTube.*

2. "Free," track 6 on Twin Atlantic, *Free*, Red Bull Records, 2011. In 2014, two years after the fall, GoPro, a company that produces action cameras, released another video showing the fall from various angles. The YouTube description reads: "GoPro was honored to be a part of this epic achievement, with seven HERO2 cameras documenting every moment. From the airless freeze of outer space, to the record-breaking free fall and momentous return to ground—see it all through Felix's eyes as captured by GoPro, and experience this incredible mission like never before. No one gets you closer than this." Go Pro, "GoPro: Red Bull Stratos—The Full Story," posted January 31, 2014 by GoPro, YouTube, https://www.youtube.com/watch?v=dYw4meRWGd4.

3. *Human to Hero*, "Daredevil Skydiver Breaks Speed of Sound," CNN, 2013.

4. Ehrenfried, *Stratonauts*, 183–5.

5. Ryan, *Magnificent Failure*, xvi.

6. Galdamez, "Exploring the Stratosphere," 235.

7. Ryan, *Magnificent Failure*, xv.

8. Ryan, *Magnificent Failure*, xv.

9. Ryan, *Magnificent Failure*, 259.

10. Wilcox, *Desert Dancing*, 137. See also Joe Mooallem, "A Journey to the Center of the World," *New York Times*, February 19, 2014, https://www.nytimes.com/2014/02/23 /magazine/a-journey-to-the-center-of-the-world.html; Jamie Lee Curtis Taete, "The History of the World Is Being Carved into Stone by an Old Man in the Desert," *VICE*, April 15, 2018, https://www.vice.com/en_us/article/ne9q4q/this-bizarre-monument-is -all-that-will-remain-of-humanity-in-4000-years.

11. See the website of the Museum of History in Granite, accessed September 6, 2019, http://www.historyingranite.org.

12. Ryan, *Magnificent Failure*, 4–5

13. Istel, "Pour Le Sport," 28.

14. Ryan, *Magnificent Failure*, 6.

15. Website of the Museum of History in Granite, accessed September 6, 2019, http:// www.historyingranite.org.

16. Ryan, *Magnificent Failure*, 3.

17. "Red Bull Refugee Challenge | NEO MAGAZIN ROYALE mit Jan Böhmermann - ZDFneo," posted December 7, 2017, by ZDF MAGAZIN ROYALE, YouTube, https://www .youtube.com/watch?v=qsf_hmBV5FQ.

18. "Falling Faster: The Surprising Leap of Felix Baumgartner," *Technical University Munich*, news release, December 14, 2017, https://www.tum.de/nc/en/about-tum/news /press-releases/details/34341/; Guerster and Walter, "Aerodynamics."

19. Barad, *Meeting the Universe Halfway*, 152, 70.

20. Barad, *Meeting the Universe Halfway*, 65.

21. Felix Baumgartner, "Liebe Facebook Freunde, Fans, Hasser, Journalisten, Politiker und Sonstige!," Facebook, January 26, 2016, https://www.facebook.com /FelixBaumgartner/photos/liebe-facebook-freunde-fans-hasser-journalisten-politiker -und-sonstigees-ist-imm/10153949158828804/.

22. See "Austria Far-Right Activist Condemned over Swastika," BBC, April 5, 2019, https://www.bbc.com/news/world-europe-47822454; Ben Quinn and Jason Wilson, "Anti-Islamic Extremist Permanently Excluded from Entering UK," *The Guardian (UK Edition)*, June 26, 2019, https://www.theguardian.com/world/2019/jun/26/anti-muslim -extremist-martin-sellner-permanently-excluded-from-entering-uk.

23. Mike Stuchbery (@MikeStuchbery_), "Martin Sellner, figurehead of Generation Identity has been *permanently* excluded from entering the United Kingdom (via GI's Telegram channel)," *Twitter*, September 16, 2019, https://twitter.com/MikeStuchbery_ /status/1143575256503918593.

24. "Verfassungsschutz stuft Identitäre als klar rechtsextremistisch ein," *Süddeutsche Zeitung*, July 1, 2019, https://www.sueddeutsche.de/politik/identitaere-bewegung -verfassungsschutz-1.4520603; "Identitäre Bewegung als rechtsextremistisch eingestuft," *Frankfurter Allgemeine Zeitung*, July, 2019, https://www.faz.net/agenturmeldungen/dpa /identitaere-bewegung-als-rechtsextremistisch-eingestuft-16279525.html.

25. Felix Baumgartner, "Ein HISTORISCHER TAG in der deutschsprachigen TV Geschichte!," Facebook, October 23, 2016, accessed May 2019.

26. Interview is available online: Hubert Patterer and Gerhard Nöhrer, "Dietrich Materschitz Im Interview: Red Bull-Chef Rechnet Mit Österreichs Flüchtlingspolitik Ab," *Kleine Zeitung*, April 8, 2017, https://www.kleinezeitung.at/steiermark/chronik/5197881 /Dietrich-Mateschitz-im-Interview_Red-BullChef-rechnet-mit.

27. Emphases are my own. The original reads as follows: "Es gibt bei allem eine kritische Masse. Am Wochenende 20.000 Wanderer und Mountainbiker im National- park Hohe Tauern, das geht. Da rücken die Gämsen halt zusammen. Wenn du aus den 20.000 aber 200.000 machst oder gar zwei Millionen, dann geht das Ganze kaputt. Wir müssen verstehen, dass nicht nur die Naturregionen endlich sind, sondern alle Ressou- rcen, Energie, Wasser, Lebensmittel, Luft, medizinische Versorgung, alles, auch die Erde selbst." Patterer and Nöhrer, "Dietrich Materschitz."

28. Alex Wellerstein, "Critical Mass," *The Nuclear Secrecy Blog*, April 10, 2015. The OED confirms 1941 as the year the term was first used. *Oxford English Dictionary*, "critical mass (*n.*)," accessed June 4, 2019, https://doi.org/10.1093/OED/7012511660.

29. Patterer and Nöhrer, "Dietrich Materschitz."

30. Felix Baumgartner, "Das ist doch mal eine Ansage!!!," Facebook, January 25, 2016, https://www.facebook.com/FelixBaumgartner/posts/das-ist-doch-mal-eine-ansage /10153947132958804/.

31. Felix Baumgartner, "Liebe Facebook Freunde." See also Baumgartner's interview with the *Frankfurter Allgemeine Zeitung*, "Alles im Leben hat ein Preisschild," *Frankfurter Allgemeine Zeitung*, accessed August 15, 2019, https://www.faz.net/aktuell/sport/mehr -sport/felix-baumgartners-leben-nach-dem-stratosphaeren-sprung-13293757-p3.html.

32. WHO, *Coronavirus Disease (COVID-19), Situation Report—121*, May 20, 2020, https://www.who.int/docs/default-source/coronaviruse/situation-reports/20200520 -covid-19-sitrep-121.pdf?sfvrsn=c4be2ec6_2.

33. WHO, *Coronavirus Disease (COVID-19), Situation Report—125*, May 24, 2020, https://www.who.int/docs/default-source/coronaviruse/situation-reports/20200524 -covid-19-sitrep-125.pdf?sfvrsn=80e7d7f0_2.

34. Felix Baumgartner, "SO SIND WIR NICHT 😊 😂😆 AT," Facebook, May 24, 2020, https://www.facebook.com/FelixBaumgartner/posts/so-sind-wir-nicht-ein-kanzler-kurz -der-keine-corona-abstandsregeln-befolgt-ein-p/10158659293863804/.

35. Felix Baumgartner, "I am always amazed how fast people react with criticism," Facebook, July 23, 2021, https://www.facebook.com/profile/100050286948315/search /?q=july%202021.

## 8. HOME OF THE MIND

1. O'Brien, "Advertising in Space," 91; Balsamello, "When You Wish," 1772–3.

2. O'Brien, "Advertising in Space," 102.

3. O'Brien, "Advertising in Space," 102.

4. 49 U.S.C. § 70102(9); cf. 103rd Cong., H.R. 2599, § 2(c)(2).

5. O'Brien, "Advertising in Space," 99–101.

6. Balsamello, "When You Wish," 1784.

7. Balsamello, "When You Wish," 1819.

8. StartRocket website, https://startrocket.me (accessed July 12, 2019). The push-back the company received led to its change of course. Acknowledging the problem of space debris, StartRocket launched a "new project," the "securing of space for a brighter tomorrow," using "foam" (instead of metal) as material for various space endeavors as it promises a "new way of space exploration." The website now informs its visitors about the importance of cleaning up space debris and introduces the "Foam Debris Catcher," a spacecraft series yet to be built and "created to help us clear the way to distant frontiers for future generations within a few years." Upon scrolling to the bottom of the page, we learn that this is only the first of many devices which makes use of the "new building material for outer space colonies." StarRocket website, https://startrocket.me/7/ (accessed August 18, 2024).

9. Steven Asarch, "Dogecoin's Cocreator Explains How the 'Parody' Currency Turned into a Billion-Dollar Movement," *Business Insider*, April 15, 2021, https://www .businessinsider.com/dogecoin-go-back-down-right-now-price-livee-billy-markus -2021-4.

10. See part I of this book.

11. Grace Dean, "Elon Musk Says That SpaceX Is Going to 'Put a Literal Dogecoin on the Literal Moon,'" *Business Insider*, April 1, 2021, https://www.businessinsider.com/elon-musk-spacex-dogecoin-moon-crypto-cryptocurrencies-2021–4.

12. Geometric Energy Corporation, "SpaceX to Launch DOGE-1 to the Moon!," *PR Newswire*, May 9, 2021, https://www.prnewswire.com/news-releases/spacex-to-launch-doge-1-to-the-moon-301287016.html.

13. See "Geometric Space Launch 1," Geometric Space, accessed June 11, 2022, https://www.geometricspace.ca/launch/1.

14. Web 3.0 refers to a new, decentralized Internet architecture built on blockchain technology and utilizing distributed ledgers that are collectively managed by its users.

15. "XI Protocol," GitBook, accessed July 19, 2024, https://xi-protocol.gitbook.io.

16. See section on space tokens: "XI Protocol," GitBook, accessed July 19, 2024, https://xi-protocol.gitbook.io.

17. Tim Sandle, "Stellar Marketing: Musk Sends Advertising into Orbit," *Digital Journal*, July 7, 2021, https://www.digitaljournal.com/tech-science/stellar-marketing-musk-sends-advertising-into-orbit/article#ixzz72AvhnR8R.

18. For a more concrete idea of how "space enhanced advertisement" is imagined in this case, see the video posted by Your Time In Space (XI Protocol) on their X page. Your Time In Space (XI Protocol), "Check out the final installment of the XI Protocol series," September 28, 2022, https://x.com/YourTimeInSpace/status/1575238486403125249.

19. See chapter 1 for context.

20. Herr Fuchs, "Dogecoin Song—To the Moon [Official]," YouTube, February 9, 2021, https://www.youtube.com/watch?v=s3NWyh8a5to

21. "New Currency, Red Bull Racing," Red Bull website, March 3, 2020, https://www.redbull.com/int-en/redbullracing/new-currency; "Red Bull Racing F1 Team Signs NFT Deal with Tezos," *Ledger Insights*, March 24, 2021, https://www.ledgerinsights.com/red-bull-racing-f1-team-signs-nft-deal-with-tezos/.

22. Rob Pope, "How to Be Superhuman," Red Bull podcast, trailer released on February 25, 2020, https://www.redbull.com/int-en/podcast-shows/red-bull-how-to-be-superhuman-podcast.

23. See website of Red Bull Air Force, accessed September 7, 2019, https://www.redbullairforce.com.

24. Stefan Müller, "Red Bull reagiert wirklich zynisch," *Die Zeit*, May 2, 2013, https://www.zeit.de/2013/19/red-bull-extremsport-marketing-helmar-buechel.

25. Martin Schauhuber and Sigi Lützow, "Gefallen für Red Bull," *Der Standard*, January 7, 2018, https://www.derstandard.at/story/2000071506906/gefallen-fuer-red-bull.

26. Elon Musk, "Making Humans a Muliplanetary Species," posted September 27, 2016, by SpaceX, YouTube, https://www.youtube.com/watch?v=H7Uyfqi_TE8.

27. Claire Grodon, "Elon Musk Says 'Probably People Will Die' in Mars Mission," *Inverse*, June 2016, https://www.inverse.com/article/16857-elon-musk-says-probably-people-will-die-in-mars-mission.

28. Redfield, "Half-Life," 797.

29. MacDonald, "Anti-Astropolitik," 610.

30. Benjamin, *Rocket Dreams*, 40.

31. O'Neill, *The High Frontier*, 27.

32. O'Neill, *The High Frontier*, 109.

33. See chapter 5 in this volume and Gandorfer, "Down and Dirty."

34. Persson, "Citizens of Mars Ltd.," 125–26.

35. David Valentin defines NewSpace as follows: "NewSpace is a neologism describing a broad range of primarily US-based entrepreneurs and advocates who, for more than 30 years, have aimed to commercialize outer space—thus far the province of state-led space programs—through enterprises as diverse as providing launches to orbit, space tourism and hotels, space-based solar power generation and delivery, moon settlement, and asteroid mining." A significant characteristic of this decentralized set of new space companies, such as SpaceX (Elon Musk), Blue Origin (Jeff Bezos), and Virgin Galactic (Richard Branson), is that these "private firms share in the enormous risks and (potential) returns of investments in space." Valentine, "Exit Strategy," 1046; Weinzierl, "Space," 180.

36. Although there is no ambiguity in the treaties, various legal scholars argue that the Outer Space Treaty only prohibits national but not private appropriation. However, at the time of the treaty's inauguration (1967), the question of private corporations venturing in space was not present. Yet, as Tennen and Sterns point out, this contention "conveniently ignores the provisions of the Moon Agreement, which expressly mention private entities in relation to the non-appropriation of celestial bodies." Sterns and Tennen, "Privateering," 2434.

37. Collins, "Efficient Allocation," 206.

38. Collins, "Efficient Allocation," 212.

39. Collins, "Efficient Allocation," 212.

40. Collins, "Efficient Allocation," 212n59 (emphasis added).

41. Collins, "Efficient Allocation," 215, 218.

42. Article 1 of the OST states: "The exploration and use of outer space, including the Moon and other celestial bodies, shall be carried out for the benefit and in the interests of all countries, irrespective of their degree of economic or scientific development, and shall be the province of all mankind." In addition, Article 11 of the Moon Treaty states that "the Moon and its natural resources are the common heritage of mankind." As Fountain, although critical of the doctrine and in favor of commercial space exploration, explains: "There are five elements generally considered central to the modern Common Heritage doctrine: 1) the area is not subject to national appropriation; 2) all states share in the management of the area; 3) the benefits derived from exploitation of resources in the area must be shared with all regardless of the level of participation; 4) the area must be dedicated to peaceful purposes; and 5) the area must be preserved for future generations." Treaty on Principles Governing the Activities of States in the Exploration and Use of Outer Space, including the Moon and Other Celestial Bodies (Outer Space Treaty), 18 UST 2410, 610 UNTS 205, 6 ILM 386 (1967), signed on January 27, 1967; Agreement Governing the Activities of States on the Moon and Other Celestial Bodies (Moon Treaty), 1363 UNTS 3, signed on December 18, 1979; Fountain, "Creating Momentum," 1759.

43. Zhang, "Extraterritorial Jurisdiction," 150.

44. Article 11, paragraph 7(d), of the Moon Treaty declares that "the Moon and its natural resources are the common heritage of mankind" and stipulates an "equitable sharing by all States Parties in the benefits derived from those resources." However, the Moon Treaty is not broadly accepted among the major space faring states—only seven states have ratified the treaty—precisely because of the nonappropriation principle and the requirement of equitable sharing. Moon Treaty, 1979; Fountain, "Creating Momentum," 1764.

45. Viikari, "Legal Regime," 2429.

46. See also Debra Werner, "Space Law Workshop Exposes Rift in Legal Community over National Authority to Sanction Space Mining," *Space News*, April 17, 2018, https://spacenews.com/space-law-workshop-exposes-rift-in-legal-community-over-national-authority-to-sanction-space-mining/.

47. These legal concepts are not only considered safe—since there are allegedly modes of existence in space that would need to be effaced for colonies to take a hold—but framed as "advancing the quality of life for all mankind." Sterns and Tennen, "Privateering," 2434; Turner, "Significance of the Frontier," 37–38.

48. See Certeau, *Everyday Life*, 127.

49. Tsing, "Natural Resources," 5100, 5102.

50. Rasmussen and Lund, "Frontier Spaces," 388.

51. A., "Arch-Anarchy," 11 (emphasis added).

52. A., "Arch-Anarchy," 11.

53. While the breaking of humanmade laws is not considered a significant issue, the author anticipates some resistance in regard to breaking the laws of nature. Here, the reader is reminded that "no legislature wrote them, no executive enforces them, and no judiciary interprets them," which means that no consequences need to be feared. A., "Arch-Anarchy," 14.

54. A., "Arch-Anarchy," 14.

55. A., "Arch-Anarchy," 17.

56. A., "Arch-Anarchy," 13.

57. A., "Arch-Anarchy," 13.

58. Jones, "Extropic Thought."

59. Leo Hickman, "Felix Baumgartner Skydive: The Key Questions Answered," *Guardian*, October 15, 2012, https://www.theguardian.com/sport/shortcuts/2012/oct/15/felix-baumgartner-skydive-key-questions-answered.

60. Red Bull, "Red Bull Stratos: Felix Gets Suited," Red Bull website, July 29, 2016, https://www.redbull.com/int-en/red-bull-stratos-felix-gets-suited.

61. Alison Kervin, "The Pursuits Interview: Felix Baumgartner," *Financial Times*, April 5, 2013, https://www.ft.com/content/86734dde-9bfa-11e2-8485-00144feabdc0.

62. Gary Morley and Olivia Yasukawa, "Felix Baumgartner: What Next for the Man Who Fell to Earth?," *CNN*, December 4, 2013, http://edition.cnn.com/2013/12/04/sport/felix-baumgartner-red-bull-skydiving/index.html.

63. The advertisement can be watched here: https://www.youtube.com/watch?v=azDUNRPT9Xs. Rupert Degas, *Red Bull (Prison Escape)—TVC 2023—Voiced by Rupert Degas*, posted by Rupert Degas | Characters & Impressions, YouTube, accessed July 14, 2025, https://www.youtube.com/watch?v=DnK_F4s_V_Y.

64. Barlow, "Declaration," 29.

65. Matt Majendie, "Stratos by the Numbers: The Key Stats Behind Felix Baumgartner's Space Jump," Red Bull, October 10, 2022, https://www.redbull.com/gb-en/stratos-space-jump-key-facts-numbers (accessed July 14, 2025); John R. Quain, "Daredevil to Plunge from Outer Space in Supersonic Suit," Fox News, April 11, 2010, archived at Internet Archive, https://web.archive.org/web/20100414232320/http://www.foxnews.com/scitech/2010/04/09/felix-baumgartner-red-bull-supersonic-suit/ (accessed July 14, 2025); Fédération Aéronautique Internationale, "10-Year Anniversary of Baumgartner's Jump from the Edge of Space," FAI, October 14, 2022, https://www.fai.org/news/10-year-anniversary-baumgartner-jump-space (accessed July 14, 2025); Guinness World Records, "Felix Baumgartner: First Person to Break Sound Barrier in Freefall," Guinness World Records, https://www.guinnessworldrecords.com/records/hall-of-fame/felix-baumgartner-first-person-to-break-sound-barrier-in-freefall (accessed July 14, 2025).

66. Kakaes, "Five Schemes," 10.

67. Article 5 of the Outer Space Treaty states: "States Parties to the Treaty shall regard astronauts as envoys of mankind in outer space and shall render to them all possible assistance in the event of accident, distress, or emergency landing on the territory of another State Party or on the high seas." Outer Space Treaty, 1967.

68. Rescue Agreement, Article 2: "If, owing to accident, distress, emergency or unintended landing, the personnel of a spacecraft land in territory under the jurisdiction of a Contracting Party, it shall immediately take all possible steps to rescue them and render them all necessary assistance." Agreement on the Rescue of Astronauts, the Return of Astronauts and the Return of Objects Launched into Space (Rescue Agreement), 19 UST 7570; 672 UNTS 119; 7 ILM 149, signed on April 22, 1968.

69. Zhang, "Extraterritorial Jurisdiction," 152.

70. See, for example, Grear, "Deconstructing Anthropos"; Naffine, "The Body Bag"; Naffine, *Law's Meaning*; Esposito, *Persons and Things*.

71. Halewood, "Law's Bodies," 1340; Grear, "Towards 'Climate Justice'?," 114.

72. Grear, *Redirecting Human Rights*, 123.

73. Grear, *Redirecting Human Rights*, 116.

74. Naffine, "The Body Bag," 83. Crucial work has been done across disciplines in revealing the exclusions, cuts, and objectifications inherent in the concept of the legal subject. Margaret Davies, for example, points to the fact that the concept of the subject (or: person) creates an internal split, creating the subject as (owning) mind and (owned) body. In looking at the gendered construction of the body, Grear points to the exclusivity of the concept, arguing that the bodies excluded from law's imaginary are those which "present an irreducible challenge to legal individualism—bodies assumed to lack definition, to be fluid, and most problematically—those bodies capable of morphing to become two bodies in one." Scholarship on the colonialist and racist histories (and presents) of the concept has further exposed its proprietary structure. See, for example chapter 5 of Weheliye, *Habeas Viscus*; Best, *The Fugitive's Properties*.

75. Grear, "Towards 'Climate Justice'?," 114.

76. Delaney, "Beyond the World," 79.

77. Redfield, "Half-Life," 796 (emphasis added).

78. Zhang, "Extraterritorial Jurisdiction," 152.

79. Helmreich, "Wave Theory Social Theory"; Jones and Helmreich, "The Shape of Epidemics."

80. Fanon, *A Dying Colonialism*, 65.

81. Already in 2021, the Federal Aviation Administration (FAA) introduced the Commercial Space Astronaut Wings Program, providing guidelines and eligibility criteria for recognizing commercial astronauts in the United States. See Federal Aviation Administration, FAA *Commercial Space Astronaut Wings Program*, Order 8800.2, Washington, D.C., U.S. Department of Transportation, July 20, 2021, https://www.faa.gov /documentLibrary/media/Order/FAA_Order_8800.2.pdf, (accessed July 8, 2024).

82. Impey, "Mars and Beyond," 106.

83. James, *Deep Space Commodities*, 10.

84. Stevens, "The Price of Air," 54.

85. Cockell, "Extraterrestrial Liberty," 18–19.

86. Persson, "Citizens of Mars Ltd.," 131.

87. Cockell, "Extraterrestrial Liberty," 31, 40.

88. Stevens, "The Price of Air," 59.

89. Stevens, "The Price of Air," 57.

90. Stevens, "The Price of Air," 61.

91. See Jones's critique of Barlow's Declaration and also the controversy around Leary's "heat death" and the role attributed to Barlow. Jones, "A Critique"; More, "Heat Death."

92. Barlow's eulogy can be found online: Fun_People Archive, "Long Live Timothy Leary," May 31, 1996, http://www.langston.com/Fun_People/1996/1996BCN.html.

93. More, "The Heat Death of Timothy Leary," 37, 39.

94. Science News Staff, "Timothy Leary's Last Trip," *Science*, April 21, 1997, https:// www.sciencemag.org/news/1997/04/timothy-learys-last-trip; David Colker, "Interring the Space Age," *LA Times*, April 22, 1997, https://www.latimes.com/archives/la-xpm -1997–04–22-me-51241-story.html.

95. Science News Staff, "Timothy Leary's Last Trip."

96. See "New Cryopreservation Technology," Alcor website, October 2005, https:// www.alcor.org/library/new-cryopreservation-technology/; Andy Greenberg, "Bitcoin's Earliest Adopter Is Cryonically Freezing His Body to See the Future," *Wired*, August 28, 2014, https://www.wired.com/2014/08/hal-finney/. See also the announcement on December 16, 2014, on Alcor's website stating that "Hal Finney Becomes Alcor's 128th Patient," https://www.alcor.org/2014/12/hal-finney-becomes-alcors-128th-patient/.

97. The alternative—namely, Alcor "Neuropreservation," that is, the preservation of the brain—costs between $80,000 and $100,000. Alcor, "FAQ," https://ftp.alcor.org /FAQs/faq06.html (accessed June 12, 2019).

98. Benjamin Ross, *The Philosophy of Transhumanism*, 13–17.

99. Gandorfer, "Down and Dirty," 370.

100. Lyall and Larsen, *Space Law*, 154.

101. Theodore W. Goodman, "To the End of the Earth: A Study of the Boundary between Earth and Space," *Journal of Space Law* 36, no. 1 (2010): 91.

102. See Lyall and Larsen, *Space Law*, 165–69. Contrary to spatialist approaches, functionalist approaches are less concerned about altitude and physical conditions and more about the purpose and activities for which an object is designed.

103. Goodman, "End of the Earth," 89.

104. Lal and Nightingale, "Where Is Space?," 4.

105. Rosenfield, "Air Space," 139.

106. Nadia Drake, "Edge of Space."

107. Lyall and Larsen, *Space Law*, 160.

108. Articles 1 and 11, paragraph 2, of the Moon Treaty, 1979.

109. Barad and Gandorfer, "Political Desirings," 51.

110. Given the revived popularity and embrace of Schmitt's work, I deem it necessary to restate that his involvement with National Socialism was neither ambiguous nor a result of temporary confusion. As mentioned, Schmitt joined the Nazi party on May 1, 1933—the same day Heidegger joined—and was appointed President of the National Socialist Jurists Association the same year. In addition, he was professor of law at the University of Berlin, significantly shaping Nazi legal thought and practice. He retained this position and remained a member of the party until May 1945. He refused denazification. I also consider it important to state these facts clearly because Schmitt's thought—that is, concepts he had already developed in the early 1920s and which he continued to present long after the declared end of the Second World War—has been increasingly used and appropriated across disciplines and political positions in North American and European academia. While this is undoubtedly a worrisome development—to say the least—it is less surprising, for this appropriation is consistent with both the constant rise of fascist and racist thought as a response to increasing poverty, exploitation, violence, and fear and the legal thought that underlies Western legal theory and its attempt to secure its power and supremacy. Indeed, as Trevor J. Barnes and Claudio Minca argue, Schmitt was caught "in between a Nazi Geopolitik [geopolitics] with its unmanageable spaces and a Nazi biopolitics of racial irrationality." Precisely because, they warn, the geographical imaginations were indelibly "stained by the biopolitical violence perpetrated by the Nazis" in the territories about which he theorized, the "spatialities of genocide must thus be read in light of contributions made by rational and enlightened academics, among whom we must include Schmitt." Theory—here legal theory—cannot be separated from the material and physical world, its potentialities, atrocities, and events, with which it becomes. Barnes and Minca, "Nazi Spatial Theory," 675, 682, 683.

111. See the epilogue of this book as well as Schmitt, *Der Nomos*; Schmitt, *Land Und Meer*. Both appeared in English translations, published by Telos Press in 2006 and 2015, respectively: Schmitt, *Nomos*; Schmitt, *Land and Sea*.

112. In this context, see also Mitchell, *Carbon Democracy*.

## 9. UNDER PRESSURE

1. Aaron and Maurice, *International Law*.

2. O'Connell, "Territorial Sea," 124–25.

3. It is worth quoting Bynkershoek's contemplations on why the limit of sight is unfit to serve as definite boundary: "For does he mean the longest possible distance a man can see from the land, and that from any land whatever, from a shore, from a citadel, from a city? As far as a man can see with the naked eye? Or with the recently invented telescope? As far as the ordinary man can see, or he that has sharp eye-sight? Surely not as far as the

keenest of sight can see, for in the ancient writers we are told people who could see all the way from Sicily to Carthage. And so this rule also is wavering and indefinite." Bynkershoek, *Cuius*, 44.

4. Skelton, *Forensic Sciences*, 126–27.

5. Kneubuehl, *Wound Ballistics*, 39.

6. O'Connell, "Territorial Sea," 124.

7. Klein, *Maritime Security*, 24.

8. The United Nations Convention on the Law of the Sea (UNCLOS) is an international treaty that provides the overarching regulatory framework for the governance of the oceans. It has 167 parties, was adopted in 1982, and entered into force in 1994. For an overview of the treaty's key points, visit the website of the United Nations: https://www.un.org/depts/los/convention_agreements/convention_overview_convention.html (accessed September 24, 2019).

9. Jones, "Lunar–Solar Rhythmpatterns," 2287.

10. "Leben Sie recht wohl und lassen Ihren Taucher je eher je lieber ersaufen. Es ist nicht übel, da ich meine Paare in das Feuer und aus dem Feuer bringe, daß Ihr Held sich das entgegengesetzte Element aussucht." Letter from Goethe to Schiller on June 10, 1797. Goethe, *Briefwechsel*, 382.

11. "O du Ausgeburt der Hölle! / Soll das ganze Haus ersaufen? / Seh ich über jede Schwelle / Doch schon Wasserströme laufen." Goethe, "The Sorcerer's Apprentice," 107; Goethe, "Der Zauberlehrling," 152.

12. "Und sie laufen! Naß und nässer / Wird's im Saal und auf den Stufen. / Welch entsetzliches Gewässer! / Herr und Meister! hör mich rufen!—/ Die ich rief, die Geister, / Werd ich nun nicht los." Goethe, "The Sorcerer's Apprentice," 107, 109; Goethe, "Der Zauberlehrling," 153.

13. "*Gehorchen! Herrschen!*—ungeheuere, schwindlichte Kluft . . . *Gehorchen und Herrschen!—Sein oder Nichtsein.*" Schiller, "Die Verschwörung," 381–82.

14. Joiner and NOAA Diving Program, *NOAA Diving Manual*, 2–9.

15. According to Vitruvius, the contractor did commit the theft. Vitruvius, *Ten Books*, 253–54.

16. Joiner and NOAA Diving Program, *NOAA Diving Manual*, 10–24.

17. The information concerning the biochemical and physical phenomena associated with diving was mainly taken from Joiner and NOAA Diving Program, *NOAA Diving Manual*.

18. "Und athmete lang und athmete tief." Schiller, "Der Taucher," 374. All English quotations from this ballad are from Schiller, "The Diver," as given on the website of the Virginia Commonwealth University created by Robert Godwin-Jones, 1994–2014, citing an anonymous 1902 translation, http://germanstories.vcu.edu/schiller/taucher_e.html (accessed August 3, 2019). I will hereafter cite it as "The Diver."

19. "Sonst wär er ins Bodenlose gefallen." Schiller, "The Diver"; Schiller, "Der Taucher," 375.

20. Goethe, *Briefwechsel*, 431.

21. "Lang lebe der König! Es freue sich, / Wer da athmet im rosigten Licht. / Da unten aber ists fürchterlich, / Und der Mensch versuche die Götter nicht." Schiller, "The Diver"; Schiller, "Der Taucher," 374.

22. This is the Greek meaning.

23. Glissant, *Poetics of Relation*, 28, 82.

24. Barad and Gandorfer, "Political Desirings," 49.

25. Feynman et al., *Lectures*, 3–11.

26. *Encyclopaedia Britannica Online*, "Air. Atmospheric Gas," last updated May 2, 2025, https://www.britannica.com/science/air.

27. Irigaray, *Forgetting of Air*, 2.

28. Although this will not be the focus of the chapter, it is not surprising that this is most explicitly pronounced in the work of Carl Schmitt. His fatal attraction and the gravity of his thought result, at least to a large extent, from his persistence in fashioning a legal theory that, on the one hand, lays claim to a specific state of matter (solids, land) and actively promotes the dismissal of another (fluids, oceans), and, on the other hand, allows sovereign power and legal force to transcend both.

29. Irigaray, *Forgetting of Air*, 12. It should be noted that the politics and materialization of sexual differences, yet another iteration of Cartesian modes of thought, are a crucial part of Irigaray's argument, according to which the rule of logos, patriarchy, and disembodied thought has dominated, and continues to dominate, Western philosophy.

30. Irigaray, *Forgetting of Air*, 7.

31. Irigaray, *Forgetting of Air*, 12.

32. Irigaray, *Forgetting of Air*, 9 (my translation).

33. Sublette and Scientists Federation of American, "High Energy Weapons."

34. Barad, "No Small Matter," G105–6.

35. "Versuchst dus noch einmal und bringst mir Kunde, / Chor Was du sahst auf des Meers tiefunterstem Grunde?" Schiller, "The Diver"; Schiller, "Der Taucher," 376.

36. Baumgartner, "Alles im Leben hat ein Preisschild," *Frankfurter Allgemeine Zeitung*, accessed August 15, 2019, https://www.faz.net/aktuell/sport/mehr-sport/felix-baumgartners-leben-nach-dem-stratosphaeren-sprung-13293757-p3.html.

37. Neumann, "Schillers Dramatische Fragmente," 35, 45. Neumann argues that it is possible to assume from Schiller's work that he had sensed a political topography that would later, in the twentieth century, be described in great detail in Schmitt's work on the alleged (legal and political) difference between land and sea.

38. Steinberg, "Sovereignty," 468.

39. Havice and Zalik, "Ocean Frontiers," 1; Steinberg, "Of Other Seas," 159–60.

40. Steinberg, "Of Other Seas," 161.

41. Steinberg, "Of Other Seas," 159.

42. Steinberg and Peters, "Wet Ontologies," 252.

43. Steinberg and Peters, "Wet Ontologies," 256, 260–61. In a different essay, Peters writes that the distinct "'hydro' materiality of the sea" and the "force of the sea" have the potential to help us address—matterphorically, we might add—not only human and nonhuman relationality but also the fact that "certain material forces result in unequal, one-sided relational compositions." Even though the conception of force by Peters differs from the understanding presented in this chapter, the importance of matter and force in thinking and understanding relations is important to emphasize. Peters, "Material Hydro-Worlds," 1242, 1253. For an insightful response to the coauthored essay on "Wet

Ontologies," drawing out political implications and histories, see Helmreich, "Seagoing Nightmares."

44. Helmreich, "Seagoing Nightmares," 1. For a more detailed account of how theory matters, see Helmreich et al., "Doing Theory."

45. Helmreich, "Seagoing Nightmares," 2 (emphasis added).

46. Helmreich, "Seagoing Nightmares," 2.

47. Lehman et al., "Turbulent Waters," 194.

48. Lehman et al., "Turbulent Waters," 192.

49. *Encyclopaedia Britannica Online*, "Water," last updated May 25, 2025, https://www.britannica.com/science/water.

50. Krebs, *Chemical Elements*, 2.

51. These questions are also asked in a current investigation conducted by the Logische Phantasie Lab titled "Movement and Organ(ization): The Criminalization of Resistance." The investigation is co-led by Lindsay Ofrias and draws from the extensive research she has conducted in the course of her doctoral work. Logische Phantasie Lab, "Movement and Organ(ization): The Criminalization of Resistance," https://lo-ph.agency/movement-and-organ. See also Ofrias and Roecker, "Organized Criminals."

52. The refusal to question the onto-epistemological assumptions about space, time, and matter (among others inherited from Western scientific and philosophical thought and by now significantly challenged) sustains legal thought and sensemaking. Examples are plenty. Delaney, for example, offers an important observation that testifies to the presence of representationalism across modes of sensemaking. He points out a movement from "discursivity of space to the discursive spatiality of law" and exposes that the liberal legal discourse makes deliberate and ample use of spatial metaphors and tropes, which are "foundationally constitutive of liberal legality as such." Metaphors, and I would add, representational devices of sensemaking more generally, allow law to "seal itself off from the world," to transcend any physical and embodied limits so that, consequently, "there is nothing that is off limits *to* law, nothing that is beyond legal signification." Put differently, metaphors and spatial imaginings are constructed as sources for power claims and legal force that "work to justify 'on-the-ground' reconfigurations." Delaney, "Beyond the World," 69–71.

53. For a more in-depth argument on matterphorics and legal concepts, see Gandorfer, "What Is Your Power?"

54. Philippopoulos-Mihalopoulos, "Lively Agency," 215.

55. Delaney, "Beyond the World," 78.

56. Delaney, "Beyond the World," 79, 80.

57. The seven nations that staked territorial claims to Antarctica were: Argentina, Australia, Chile, France, New Zealand, Norway, and the United Kingdom. The sovereignty claims of Argentina, Chile, and the UK overlap. In addition, Russia reserved the right to make future territorial claims to Antarctica. See Hund, "Antarctic Territorial Claims," 49.

58. The Antarctic Treaty is the core document of the Antarctic Treaty System (ATS). However, the ATS also includes the ARCM Recommendations and Measures, the Convention of the Conservation of Antarctic Seals (1978), the Convention on the Conservation of Antarctic Marine Living Resources (1982), the Protocol on Environmental Protection to the Antarctic Treaty (1998), and the Agreement on the Conservation of Albatrosses

and Petrels. Lukin, "Antarctic Treaty System," 53; Antarctic Treaty, 12 UST 794; 402 UNTS 71; 19 ILM 860 (1980), signed on December 1, 1959, articles I–IV.

59. Article 3, paragraph 1 of the Protocol on Environmental Protection to the Antarctic Treaty (Madrid Protocol), 30 ILM 1455, signed on October 4, 1991.

60. Article 7 of the Madrid Protocol, 1991.

61. Collis, "Territories," 94.

62. Soucek, "The Polar Regions," 280; Collis, "Territories," 293, 294.

63. Joyner, "Ice-Covered Regions," 216–17.

64. Erika, "Two Poles," 32–33.

65. Erika, "Two Poles," 35; Joyner, "Ice-Covered Regions," 217.

66. Hund, "Antarctic, Definitions Of," 54–55.

67. Steinberg, "Of Other Seas," 163; Steinberg and Peters, "Wet Ontologies," 260.

68. Joyner, "Ice-Covered Regions," 242.

69. Joyner, "Ice-Covered Regions," 242.

70. Joyner, "Ice-Covered Regions," 230.

71. Joyner, "Ice-Covered Regions," 238.

72. Joyner, "Ice-Covered Regions," 242; Lewis, "Iceberg Harvesting," 440.

73. Povinelli et al., "Mattering-Forth," 304.

74. See Aston and Davies, "Grounds."

75. Cullinan, *Wild Law*, 72.

76. Cullinan, *Wild Law*, 142.

77. Cullinan, *Wild Law*, 78, 79.

78. Cullinan, *Wild Law*, 48.

79. Cullinan, *Wild Law*, 128. This is not to be confused with Derrida's understanding of a text. Cullinan refers to the physical world and puts "primary texts," which is the term for a direct resource of information, such as an original text and a first-hand account, under quotation marks to emphasize the crucial role of the physical world for legal philosophy and jurisprudence.

80. Cullinan, *Wild Law*, 174.

81. Feynman et al., *Lectures*, 1, 1–2.

82. Feynman et al., *Lectures*, 2–3.

83. Philippopoulos-Mihalopoulos, *Spatial Justice*, 25. For a creative and matterphorical engagement with questions of law and legal meaning, see Gandorfer and Ayub, "Matterphorical"; Goodrich et al., *Law and Literature*.

84. Philippopoulos-Mihalopoulos, *Spatial Justice*, 47.

85. Barad, "No Small Matter," G111.

86. Barad, "Invertebrate Visions," 229.

87. Barad, "Quantum Entanglements," 261.

88. "Den Jüngling bringt keines wieder." Schiller, "The Diver"; Schiller, "Der Taucher," 376.

89. Joiner and NOAA Diving Program, *NOAA Diving Manual*, 2 16.

90. "Denn unter mir lags noch, Bergetief, / In purpurner Finsterniß da, / Und obs hier dem Ohre gleich ewig schlief." Schiller, "The Diver" (adjusted: "forever"); Schiller, "Der Taucher," 375.

91. Dziak et al., "Ambient Sound," 190; Heffernan, "Deep-Sea Dilemma," 488.

92. Dziak et al., "Ambient Sound," 186. For an in-depth view on the interconnected-ness and dynamic interactions between sound, ocean, and air, highlighting their roles in environmental, technological, and biological contexts, see the following article in the *Matterphorical* special issue of *Theory and Event*: Ganchrow, "Earth-Bound Sound."

93. See interview with James Cameron: Josh Rottenberg, "James Cameron on His Deep-Sea Doc and Managing a Fathomless Project," *Fast Company* (blog), August 6, 2014, https://www.fastcompany.com/3033243/james-cameron-on-his-deep-sea-doc-and -managing-a-fathomless-project.

94. William J. Broad, "So You Think You Dove the Deepest? James Cameron Doesn't," *New York Times*, September 16, 2019, https://www.nytimes.com/2019/09/16/science /ocean-sea-challenger-exploration-james-cameron.html.

95. Otto, *Unterwasser-Literatur*, 197.

96. Havice and Zalik, "Ocean Frontiers," 1.

97. Havice and Zalik, "Ocean Frontiers," 10.

98. Mansfield, "Neoliberalism," 324–25.

99. Casson, "Greenpeace," 3.

100. Casson, "Greenpeace," 6.

101. Deep Sea Mining Campaign, London Mining Network, and Mining Watch Canada, "Why the Rush?," 1.

102. Deep Sea Mining Campaign, London Mining Network, and Mining Watch Canada, "Why the Rush?," 4.

103. Fountain, "Creating Momentum in Space," 1769.

104. Rasmussen and Lund, "Reconfiguring Frontier Spaces," 391.

105. Weizman, *Forensic Architecture*, 233.

106. Weizman, *Forensic Architecture*, 228.

107. Weizman and Sheikh, *The Conflict Shoreline*, 8.

108. Weizman and Sheikh, *The Conflict Shoreline*, 9.

109. Koskenniemi, *Apology*, 16.

110. Haraway, *Staying with the Trouble*, 3; Massumi, *Postcapitalist Manifesto*, 69.


10. MATTERPHORICS OF LIFE

1. Latour, *Facing Gaia*, 226, 77.

2. Latour, *Facing Gaia*, 228.

3. Latour's argument stating that Schmitt, looking at the earth, "sees the matrix of a possible regime of law" and that "someone who is ignorant of nature to this extent is exactly what we need!" relies on a questionable reading and misunderstands Schmitt's conception of the relation between nature, matter, and law. Latour selectively uses con-cepts from Schmitt's *The Nomos of the Earth in the International Law of the Jus Publicum Europaeum*, which is consistent with Schmitt's earlier work. In *Nomos* and *Land and Sea*, Schmitt elaborates his Eurocentric legal vision rooted in land as territory, emphasiz-ing the oppositional relationship between land and sea. Schmitt defines man as a "land being" (Landwesen), excluding those without this spatial relationship to earth and land from humanity. Joshua Derman notes that, for Schmitt, "man is a creature of the Earth, Britons are 'children of the sea' and the Jews are utterly foreign to the land, ergo both are

artfremd in relation to the human species." Schmitt's approach naturalizes the German and European legal order, creating an unquestionable origin of legal force that transcends into the immaterial. Confidently, he asserts that "world history" is solely "the history of conflicts between land and maritime powers," with man—meaning those included in the category of the human—transcending his ties to matter through history and intellectual superiority. Indeed, man, Schmitt continues, "knows not only the possibility of being born, but also that of rebirth," as, unlike the animal or the plant, he can, "by means of his mind, unwavering observations and analytical conclusions, and by means of decision transcend into a new existence." Latour, *Facing Gaia*, 230; Derman, "Carl Schmitt," 189; Schmitt, *Land Und Meer*, I, 2–15, 16.

4. Haraway, *Staying with the Trouble*, 12.

5. Haraway, *Staying with the Trouble*, 43 (emphasis added).

6. The right to life is considered the most fundamental of all human rights and as such is recognized explicitly in virtually all international human rights declarations and treaties and in many national constitutions. However, it is also one that leaves room for interpretation and is far from uncontested. It is, for example "affirmed in several specific contexts, where international law prohibits use of the death penalty, the intentional killing of civilians in armed conflict, and the perpetration of genocide." However, "in other areas germane to the right to life, such as abortion and euthanasia, it has proved impossible to reach any international consensus given vastly different attitudes based upon geography, culture, and social development." Right to life, therefore, denotes a (legal) right and is not to be confused with its many politically or morally motivated appropriations, as prevalent, for example, in the rhetoric around abortion in the United States. With the phrase *right to (a) life*, I wish to go conceptually further by taking into consideration that life, even in the legal context, is considered as something the subject can (or cannot) possess. It is, as such *a* life, definable, with recognizable contours and features. For a more detailed elaboration on the right to life and its history, see Schabas, "Right to Life."

7. Crutzen, "Geology of Mankind," 23.

8. Crist, "Nomenclature," 16–17.

9. Povinelli, *Geontologies*, 9.

10. Moore, "Cheap Nature," 82–83.

11. Povinelli, *Geontologies*, 12.

12. Haraway, *Staying with the Trouble*, 39.

13. Fleurke et al., *Constitutionalizing in the Anthropocene*, 22.

14. Grear, "Deconstructing Anthropos," 231, 233.

15. Helmreich, *Limits of Life*, 1.

16. Jasanoff, "Taking Life," 181. In this essay, Jasanoff explores crucial US cases (as well as a Canadian case) regarding the patenting of life and life forms and draws a line from these "takings of life" to those of real property. In doing so, she emphasizes the underlying economical workings of law and, as a consequence, the linkage between the establishment of ownership and the capitalist notion of reviving idle nature—be it land or organisms. She concludes that "whether nature resides in inanimate land or in living things, what the law rewards is the act of economic agency that takes something that was fixed, embedded, and immovable and makes it specific, dynamic, and commercially value-laden. In short, *lively*."

17. Diamond v. Chakrabarty, 447 U.S. 303 (1980).

18. Jasanoff, "Taking Life," 167, 172.

19. Jasanoff, "Taking Life," 181.

20. Garforth, "Life," 47.

21. Shiva, *Who Really Feeds the World*, 70–71; Shiva, *Biopiracy*, xviii.

22. Garforth, "Life," 42.

23. Christopher M. Holman, a US legal scholar and proponent of the said extension of copyright protection, describes the "striking" analogy as follows: "A genetic sequence provides a series of instructions directing a living cell to perform functions dictated by the instructions. Genetic engineering permits a human to dictate these instructions. Like a computer program, a genetic sequence can be expressed in a format directly interpretable by a human, albeit instead of a series of zeros and ones, it is a sequence of A, T, C and G's, representing the four primary nucleotides that make up DNA." Holman, "Copyright for Engineered DNA," 713.

24. Roosth, *Synthetic*, 12.

25. Roosth, *Synthetic*, 83.

26. Jasanoff, "Taking Life," 162.

27. Delaney, "Lively Agency," 216–17.

28. Philippopoulos-Mihalopoulos, *Spatial Justice*, 202.

29. Cullinan, "Wild Law," 72.

30. Povinelli, *Geontologies*, 9.

## 11. WHEN THEORY CRASHES (INTO) LIFE

1. Deleuze, *Pure Immanence*.

2. Barad, "On Touching," 208.

3. Braidotti, "Animals," 527.

4. Whitehead, *Reason*, 4.

5. Haraway, "Situated Knowledges," 580.

6. Taylor and Dewsbury, "Metaphor Use," 1.

7. Taylor and Dewsbury, "Metaphor Use," 2; Nelkin, "Molecular Metaphors," 556; Keller, *Making Sense*, 119.

8. Jasanoff, *Science*, 143, 179. The sociologist of science, Dorothy Nelkin, remarked in 2001 that the circulating genetic metaphors can be classified as essentialist (genes as essence of personal identity), religious (genes as sacred), fatalistic (genes as destiny), and commercial metaphors (genes as commodities). Nelkin and Roosth, among others, emphasize the Judeo-Christian thinking underlying various metaphors, among them DNA and the genome as the book of life, and, even more generally, the idea of life as a text that can be read, decoded, and, ultimately, even rewritten. What is more, the flawed understanding of DNA conveyed by metaphor as "relatively independent of the body, giving the body life, power and true identity," evokes a conception of DNA as "an invisible entity," not only immortal but also "containing in it everything needed to bring the body back." DNA is, in other words, not life but "a biological entity, a text without context, data without dimension." Nelkin, "Molecular Metaphors," 557, 559. Roosth, *Synthetic*, 81.

9. Ball, "A Synthetic Creation Story."

10. Ball, "A Synthetic Creation Story."

11. Of course, scientific theories are different from legal or literary theories. In addition to the empirical or mathematical proof, the inherent instability of representational systems, such as language, adds to the difficulty of articulating a theory of life in the sciences. However, my point is not to compare theories but to show that the question of the relationship between life and theory runs through various disciplines and fields, urging us, from wherever we decide to take that question up, to address its complexity and the matterphorical concept at stake.

12. Keller, *Making Sense*, 3.

13. Helmreich, *Limits of Life*, 11.

14. Keller, *Making Sense*, 2.

15. Keller, *Making Sense*, 1–2.

16. Jasanoff, *Science*, 3.

17. Povinelli, *Geontologies*, 14.

18. Povinelli, *Geontologies*, 38.

19. Povinelli, *Geontologies*, 43.

20. Povinelli, *Geontologies*, 14.

21. Povinelli, *Geontologies*, 176. Povinelli argues that geontology "is not a crisis of life (*bios*) and death (*thanatos*) at a species level (extinction), or merely a crisis between Life (*bios*) and Nonlife (*geos, meteoros*). Geontopower is a mode of late *liberal* governance." Povinelli, *Geontologies*, 16.

22. Povinelli, *Geontologies*, 36–37.

23. Kimberly TallBear, "Life/Not Life," 180.

24. Kimberly TallBear, "Life/Not Life," 182.

25. Kimberly TallBear, "Life/Not Life," 194.

26. Roosth, "Life, Not Itself," 59.

27. Roosth, "Life, Not Itself," 59.

28. Roosth, "Life, Not Itself," 75.

29. Roosth, "Life, Not Itself," 60.

30. Roosth, "Life, Not Itself," 75.

31. Roosth, "Life, Not Itself," 60.

32. Roosth, *Synthetic*, 9.

33. Roosth, *Synthetic*, 8, 38.

34. Roosth, *Synthetic*, 155.

35. Helmreich, *Limits of Life*, xiv.

36. Helmreich, *Limits of Life*, xiv.

37. Helmreich, *Limits of Life*, xvi, 4.

38. Helmreich, *Limits of Life*, xvii.

39. Feynman et al., *Lectures*, 1–2.

40. Butler, *Frames of War*, 3.

41. Butler, *Frames of War*, 7.

42. Delaney, "Afterword," 214.

43. Naffine, *Law's Meaning*, 48.

44. Delaney, "Afterword," 217.

45. Davies, *Property*, 14.

46. Naffine, *Law's Meaning*, 5; Jasanoff, "Taking Life," 162.

47. Jasanoff, "Rewriting Life," 288.

48. Coombe, "Properties of Culture," 251. See also Carolan, "Patent Law," 760.

49. James, *Shamans*, 49, x.

50. Jaszi and Woodmansee, *Authorship*, 13.

## 12. THE RIGHT TO NARRATE (A) LIFE

1. Derrida, "The Law of Genre," 56.

2. Slaughter, *Human Rights*, 11.

3. Davies, *Law Unlimited*, 43.

4. Deleuze and Guattari, *A Thousand Plateaus*, 70.

5. Barad, *Meeting the Universe Halfway*, 329.

6. Barad, "Invertebrate Visions," 234.

7. Esposito, "The Dispositif," 23.

8. Barad, *Meeting the Universe Halfway*, 233. In regard to legal theory and law, these cuts are performed arbitrarily but not without a pattern that indicates the underlying assumptions and essentialisms. According to feminist legal scholar Naffine, the casts from which legal personhood are drawn include the person as (1) "a legal rights-and-duty bearer," (2) a "human being (a biological species use)," and the (3) "moral person, as a being of inherent value, but that value may derive from different sources, such as the capacity for reason (typically a secular humanist usage) or from the supposed human spirit (a religious usage)." What is more, Naffine identifies two influential schools of thought that have played a crucial role in sustaining this anthropocentrism and essentialism; namely, secular rationalists and conservative Christians. Roberto Esposito confirms this point by stating that it is the secular and the Catholic that underlies the concept of the person in law. Naffine, *Law's Meaning*, 11; Naffine, "Legal Personality," 69; Esposito, *Third Person*, 5.

9. Jasanoff, "Taking Life," 65; Metzler, "Human Life," 296. Indeed, ontological surgeries play a major role in biotechnological research; namely, as means to extract "bio-objects" from (human) life and life forms in order to then decide "how to deal with them in normative terms, and who to entrust with their oversight." The scientific and normative operations already determine on the molecular level what "life" means, and thereby qualify what is rendered a life form and what is reduced to a "bio-object"—created in a lab and through intervening in and objectifying life, often stored, circulated, and exchanged. Ingrid Metzler and Andrew Webster, "Bio-Objects," 649.

10. Naffine, "Legal Personality," 73.

11. Esposito, "The Dispositif," 18.

12. Esposito, *Third Person*, 4.

13. "Molecular scissors," and often also "molecular scalpel," is the metaphor (in this case a *matterphor* as it does cut, yet not as claimed) used to describe CRISPR gene (genome) editing. For a more thorough analysis of the metaphor, as well as other metaphors used to describe gene editing, see, for example, Brigitte Nerlich, "Gene Editing, Metaphors and Responsible Language Use," *Making Science Public*, December 11, 2015, http://blogs .nottingham.ac.uk/makingsciencepublic/2015/12/11/59072/.

14. Fox, "Foundations," 65.

15. Fox, "Foundations," 64.

16. Fox, "Foundations," 64–65.

17. Deleuze, *Difference and Repetition*, 276.

18. Smith and Watson, *Reading Autobiography*, 222.

19. Whitehead, *Science* , 202.

20. Re A (Conjoined Twins), 2 WLR 480 (2001).

21. Harris, "Human Beings," 235.

22. Harris, "Human Beings," 233.

23. Harris, "Human Beings," 234.

24. Harris, "Immortal Ethics," 528.

25. Harris, "Immortal Ethics," 528.

26. O'Connor, "What's Wrong with Death?," 23, 27, 28.

27. In an interview with the *Guardian* on gene editing of human embryos, Harris states that it is a "duty" and a "moral obligation" to "create the best possible child." Ultimately, his rational is space colonialization, arguing that eventually we will have to "escape both beyond our fragile planet and our fragile nature." Gene editing and human enhancements are, he argues, one way to facilitate this aim. John Harris, "Why Human Gene Editing Must Not Be Stopped." *Guardian* (UK edition), December 2, 2015, https://www.theguardian.com/science/2015/dec/02/why-human-gene-editing-must-not-be-stopped. In regard to eugenics and transhumanism, see also footnote 35 (chapter 5).

28. For an analysis of Harris's transhumanist logic, see Sparrow, "Eugenics."

29. Metzler, "Human Life," 296.

30. Naffine, *Law's Meaning*, 164–65.

31. Naffine, *Law's Meaning*, 164–65.

32. In the course of this chapter, the terms *subject*, *self*, and *person* are not used synonymously but nevertheless interchangeably. Each term—subject, person, self—has its histories and genealogies, its connotations and counterparts. What ties all three terms together, and what contributes to their mingling and complicity across disciplines and genealogies, is their root in a dominant notion of the subject, person, and self as Anthropos. It is at this encounter, in that context, at that specific point, that the three terms merge. The different uses of the terms across literature as well as in the course of this chapter signal their trajectories, their encounters, and their complicities in constructing a mode of thought and knowledge production that is highly exclusive, oppressive, and, potentially, although not in evert instance and application, violent.

33. See, for example, Grear, *Redirecting Human Rights*; Naffine, *Law's Meaning*; Naffine, "The Body Bag."

34. Cetacean Community v. Bush, 386 F.3d 1169 (2004); Naruto v. Slater, 888 F.3d 418 (2018).

35. Angela Naimou, for example, shows how arguments for an uncritical expansion of legal personhood to animals or fetuses are often driven by analogies to the legacy of slavery and abolition. These arguments mainly rely on the universalisation of the "excluded." The assumption that the simple inclusion of what has been denied a right to life into a premade legal framework that still operates by means of the same exclusionary device that excluded those lives in the first place creates an illusion of formal equality while forgoing

the challenge of questioning the history of legal personhood and the complexities of the relation between life and law. And as Esposito concludes, the failure of human rights and their inability to restore the broken relation between law and life is not despite but because of the ideology at heart of legal personhood. Thus, it is precisely concept of the person and its extension—that is, the fact that "we have never really moved out of it"—that constitutes the inability to join life and rights. What is at stake is therefore not a more inclusive account of legal personhood but a conceptually different concept of how law and life relate to each other. Naimou, *Salvage Work*, 216; Esposito, *Third Person*, 5.

36. Cullinan, *Wild Law*, 48.

37. Povinelli, *Geontologies*, 42. Similarly, Michael Marder describes, in a co-authored book with Luce Irigaray, the impossibility of holding on to a notion of life that depends on a bounded body, isolated from the ecosystems that sustain its functions, matterphorically with reference to the lungs: "After all, with every breath we take, we expose our lungs to the outside world, regardless of all the barriers we have erected between the environment and ourselves." Irigaray and Marder, *Through Vegetal Being*, 130.

38. Herbrechter, "Narrating," 4. Another emerging field that challenges and dismantles Western traditions of autobiography, life writing, and life narration is postcolonial life writing. See especially Whitlock, *Postcolonial Life Narratives*; Moore-Gilbert, *Postcolonial Life-Writing*.

39. Herbrechter, "Narrating," 8.

40. Derrida, "Law of Genre," 56.

41. Slaughter, *Human Rights*, 4, 7.

42. Ryan, *Politics and Culture*, 158.

43. Smith and Watson, *De/Colonizing the Subject*, xviii.

44. Lejeune and Eakin, *On Autobiography*, 29.

45. See, for example, Anderson, *Autobiography*, 3; Straight, *Autobiography*, 9–10.

46. Smith and Watson, *De/Colonizing the Subject*, vii.

47. Derrida, *The Animal*, 56.

48. Derrida, *The Animal*, 57.

49. Derrida, *The Animal*, 24.

50. Straight, *Autobiography*, 13.

51. Straight, *Autobiography*, 13.

52. This is not to say that the concept of the contract is not already problematic and exclusive enough. Claiming that autobiography is a contractual genre—which implies, in the simplest form, the idea two legal subjects entering into an exchange relation—evokes the (liberal) idea of formal equality and thus the erasure of difference. It was especially feminist legal thought that exposed the dangers of the autonomous legal subject; that is, the "legal preoccupation with the adult, rational, self-governing, rights-enforcing individual, which has rendered exceptional and odd those who are weak and vulnerable: those who cannot achieve this rugged degree of physical self-sufficiency." See Naffine, *Law's Meaning*, 155.

53. Derrida, *The Animal*, 88.

54. Derrida, *The Animal*, 87–88.

55. Derrida, "Eating Well," 106.

56. Derrida, "Eating Well," 108–9.

57. Derrida, "Eating Well," 114.

58. Alaimo, *Bodily Natures*, 89, 95. Alaimo uses the terms *autobiography* and *memoirs* interchangeably.

59. Weil, *Autobiography*, 93.

60. Huff, "After Auto," 279.

61. Braidotti, *The Posthuman*, 49.

62. Deleuze, *Difference and Repetition*, 276.

63. Povinelli, "On Biopolitics."

64. Povinelli, *Geontologies*, 9.

65. Ball, "Synthetic Creation Story."

66. Naffine, *Law's Meaning*, 16.

67. Grear, *Redirecting Human Rights*, 115; Grear, "Deconstructing Anthropos," 239.

68. Hardack, "New and Improved," 46.

69. Hardack, "New and Improved," 37.

70. Hardack, "New and Improved," 37.

71. Hardack, "New and Improved," 60.

72. Hardack, "New and Improved," 41.

73. Hardack, "New and Improved," 37.

74. Hardack, "New and Improved," 38, 46.

75. The legal scholar Peter Halewood shows in great detail how the Enlightenment-informed production of the subject is shaped by the Cartesian mind/body dichotomy, which yields an understanding of persons, physical bodies, and property that relies heavily on the radical dichotomy between subject and object. Grear takes Halewoods arguments further and situates them in the discourse around the Anthropocene. See Grear, "Deconstructing Anthropos," 236, 237; Grear, "Towards 'Climate Justice'?," 112; Halewood, "Law's Bodies."

76. For a rather early proof of Bostrom's involvement in the discussions that took place via email, see http://extropians.weidai.com/extropians.4Q97/author.html (accessed May 2017).

77. Bostrom, *Superintelligence*, 168

78. Bostrom, *Superintelligence*, 126.

79. Cullinan, *Wild Law*, 67.

80. Bell, "Wisdomism," 23.

81. As discussed in part I and further elaborated elsewhere, transhumanist visions of future law often aim to actualize the Cartesian mind/body dualism. It is important to note that the potential risk posed by transhumanist legal thought does not stem from technology itself but rather from its philosophical underpinnings, which are historically connected to imperialism, colonialism, and the construction of the sentient, proprietary subject, endowed with language and reason. See Gandorfer, "Embodied Critique."

82. Povinelli, *Geontologies*, 9.

## 13. MATTERS OF INDETERMINACY

1. De Man, "Autobiography as De-Facement," 919.

2. Derrida, "Eating Well," 117.

3. Goodrich, *Law and Literature*, 27.

4. Philippopoulos-Mihalopoulos, *Spatial Justice*, 3.

5. Philippopoulos-Mihalopoulos, "Critical Autopoiesis," 410.

6. Philippopoulos-Mihalopoulos, "Critical Autopoiesis," 410.

7. Philippopoulos-Mihalopoulos, *Spatial Justice*, 75. Philippopoulos-Mihalopoulos follows a Spinozian/Deleuzian understanding of bodies, which means that bodies are "human, nonhuman, technological, natural, immaterial, material, elemental, systemic." Philippopoulos-Mihalopoulos, *Spatial Justice*, 5.

8. Philippopoulos-Mihalopoulos, "To Have to Do with Law," 480.

9. Goodrich, "Transhumusians," 126.

10. Pottage, "Subject of Law," 1199.

11. Povinelli, *Geontologies*, 69, 157.

12. Povinelli, *Gaia and Ground*, 130; Barad and Gandorfer, "Political Desirings," 26.

13. Povinelli et al., "Mattering-Forth," 302–3.

14. Srinivasan, *Network State*, 239, 259.

15. Williams, *Race and Rights*, 165.

16. Deleuze and Guattari, *A Thousand Plateaus*, 25.

17. Povinelli, *Gaia and Ground*, 121.

18. Povinelli, *Gaia and Ground*, 121.

19. Povinelli et al., "Mattering-Forth," 304.

20. See Gandorfer, "Down and Dirty."

21. Williams, *Race and Rights*, 163.

22. Williams, "Skittles as Matterphor"; Fanon, *A Dying Colonialism*, 65. On air pollution, see, for instance, United States Environmental Protection Agency, *Policy Assessment*; Li et al., "Air Pollution."

23. Williams, *Race and Rights*, 165.

24. Williams, *Race and Rights*, 165.

25. Deleuze and Parnet, *Dialogues II*, 51.

26. Deleuze and Parnet, *Dialogues II*, 50.

27. Deleuze and Parnet, *Dialogues II*, 43–44.

28. Deleuze and Parnet, *Dialogues II*, 52.

29. Deleuze and Parnet, *Dialogues II*, 52.

EPILOGUE

1. World Health Organization, "Air Pollution," accessed July 19, 2024, https://www.who.int/health-topics/air-pollution#tab=tab_1.

2. This differentiation refers to Spinoza's understanding of power as outlined in his *Ethics*, as well as elaborations on it by Braidotti. Spinoza, *Ethics*; Braidotti, "Theoretical Framework."

3. Ganchrow, "Chlorophyll-Ocean-Iron-Breath."

4. Foucault, "Preface," xiv.

5. Deleuze, *Difference and Repetition*, 1.

6. Difference for Deleuze is an ongoing becoming and thus affirmative; it is being and nonbeing, or, as he suggests "?-being." Deleuze, *Difference and Repetition*, 74.

7. Gandorfer, "Breathing Law," 172.

8. Deleuze, *Difference and Repetition*, 57.

9. Deleuze and Guattari, *A Thousand Plateaus,* 535. Quoted in Gandorfer, "Breathing Law," 171.

10. Williams, *Race and Rights*, 163–65.

11. Deleuze, *Difference and Repetition*, 1.

12. Deleuze, *Difference and Repetition*, 2, 3, 24.

13. Foucault, "Preface," xiv.

14. Massumi, *Postcapitalist Manifesto*, 132.

15. Massumi, *Postcapitalist Manifesto*, 69.

16. Foucault, "Preface," ix. For a more detailed account of how Foucault's caution matters in regard to new digital technologies and political experiments, see Gandorfer, "Down and Dirty."

17. Povinelli et al., "Mattering-Forth," 304.

18. Deleuze and Parnet, *Dialogues II*, 41.

19. Goodrich, "Critique of Comparative Law," 142.

# Bibliography

A. "Arch-Anarchy." *Extropy: Vaccine for Future Shock*, Winter 1990, 11–18.

Aaron, X. Fellmeth, and Horwitz Maurice. *Guide to Latin in International Law*. Oxford: Oxford University Press, 2009.

Alaimo, Stacy. *Bodily Natures: Science, Environment, and the Material Self*. Bloomington: Indiana University Press, 2010.

Alford-Jones, Kelsey. *Should the Inter-American Development Bank Fund Honduras to Implement Controversial Special Economic Zones?* Washington, DC: Center for International Environmental Law, 2017.

Anderson, Linda R. *Autobiography*. London: Routledge, 2011.

Arendt, Hannah. *Eichmann in Jerusalem: A Report on the Banality of Evil*. New York: Penguin, 2006.

Aristotle. *On the Art of Poetry*. Translated by Ingram Bywater. Oxford: Clarendon Press, 1909.

Aston, Rhys, and Margaret Davies. "Grounds." In *Research Handbook on Law and Literature*, edited by Peter Goodrich, Daniela Gandorfer, and Cecilia Gebruers. Cheltenham, UK: Edward Elgar, 2022.

Ball, Philip. "A Synthetic Creation Story." *Nature*, May 24, 2010.

Balsamello, F.J. "When You Wish Upon a Falling Billboard: Advertising in an Age of Space Tourism." *Georgetown Law Journal* 98, no. 6 (2010): 1769–822.

Barad, Karen. "After the End of the World: Entangled Nuclear Colonialisms, Matters of Force, and the Material Force of Justice." *Theory and Event* 22, no. 3 (2019): 524–50.

Barad, Karen. "Invertebrate Visions: Diffractions of the Brittlestar." In *The Multispecies Salon*, edited by Eben Kirksey. Durham, NC: Duke University Press, 2014.

Barad, Karen. *Meeting the Universe Halfway: Quantum Physics and the Entanglement of Matter and Meaning*. Durham, NC: Duke University Press, 2007.

Barad, Karen. "Meeting the Universe Halfway: Realism and Social Contructivism without Contradiction." *Feminism, Science, and the Philosophy of Science*, no. 256 (1996): 161–94.

Barad, Karen. "No Small Matter: Mushroom Clouds, Ecologies of Nothingness, and Strange Topologies of Spacetimemattering." In *Arts of Living on a Damaged Planet:*

*Ghosts and Monsters of the Anthropocene*, edited by Anna Lowenhaupt Tsing, Heather Anne Swanson, Elaine Gan, and Nils Bubandt. Minneapolis: University of Minnesota Press, 2017.

Barad, Karen. "On Touching: The Inhuman That Therefore I Am." *differences: A Journal of Feminist Cultural Studies* 23, no. 3 (2012): 206–23.

Barad, Karen. "Posthumanist Performativity: Toward an Understanding of How Matter Comes to Matter." *Signs* 28, no. 3 (2003): 801–31.

Barad, Karen. "Quantum Entanglements and Hauntological Relations of Inheritance: Dis/Continuities, Spacetime Enfoldings, and Justice-to-Come." *Derrida Today* 3, no. 2 (2010): 240–68.

Barad, Karen. *What Is the Measure of Nothingness? Infinity, Virtuality, Justice/ Was Ist Das Maß Des Nichts? Unendlichkeit, Virtualität, Gerechtigkeit*. Berlin: Hatje Cantz, 2012.

Barad, Karen, and Daniela Gandorfer. "Political Desirings: Yearnings for Mattering (,) Differently." *Theory and Event* 24, no. 1 (2021): 14–66.

Barad, Karen and Daniela Gandorfer. "Reading Force(s)." Introduction by Zulaikha Ayub. Virtual lecture, December 2018, posted July 15, 2019, by Reading Matters, YouTube. https://www.youtube.com/watch?v=rFmeXJR4Clk.

Barlow, John Perry. "A Declaration of the Independence of Cyberspace." In *Crypto Anarchy, Cyberstates, and Pirate Utopias*, edited by Peter Ludlow. Cambridge, MA: MIT Press, 2001.

Barnes, Trevor J., and Claudio Minca. "Nazi Spatial Theory: The Dark Geographies of Carl Schmitt and Walter Christaller." *Annals of the Association of American Geographers* 103, no. 3 (2013): 669–87.

Bell, Thomas W. "Extropia. A Home for Our Hopes." *Extropy: Vaccine for Future Shock*, Winter 1991–2, 35–41.

Bell, Thomas W. "Wisdomism, a Moral Theory for the Age of Information." *Extropy: Vaccine for Future Shock*, Winter 1989, 22–28.

Bell, Tom W. *Your Next Government? From the Nation State to Stateless Nations*. Cambridge: Cambridge University Press, 2017.

Benjamin, Marina. *Rocket Dreams: How the Space Age Shaped Our Vision of a World Beyond*. New York: Free Press, 2003.

Bernstein, Anya. *The Future of Immortality: Remaking Life and Death in Contemporary Russia*. Princeton, NJ: Princeton University Press, 2019.

Best, Stephen Michael. *The Fugitive's Properties: Law and the Poetics of Possession*. Chicago: University of Chicago Press, 2004.

Bignall, Simone, and Daryle Rigney. "Indigeneity, Posthumanism and Nomad Thought: Transforming Colonial Ecologies." In *Posthuman Ecologies: Complexity and Process After Deleuze*, edited by Rosi Braidotti and Simone Bignall. New York: Rowman and Littlefield International, 2019.

Bostrom, Nick. *Superintelligence: Paths, Dangers, Strategies*. Oxford: Oxford University Press, 2014.

Bostrom, Nick. "The Transhumanist FAQ: A General Introduction: Version 2.1." World Transhumanist Association, 2003. https://nickbostrom.com/views/transhumanist.pdf.

Boyle, James. *Shamans, Software, and Spleens: Law and the Construction of the Information Society.* Cambridge, MA: Harvard University Press, 1997.

Braddock, Jeremy. "Race: Tradition and Archive in the Harlem Renaissance." In *A Handbook of Modernism Studies*, edited by Jean-Michel Rabaté. Chichester: Wiley-Blackwell, 2016.

Braidotti, Rosi. "Animals, Anomalies, and Inorganic Others." *PMLA* 124, no. 2 (2009): 526–32.

Braidotti, Rosi. "Four Theses on Posthumanist Feminism." In *Anthropocene Feminism*, edited by Richard A. Grusin, 21–48. Minneapolis: University of Minnesota Press, 2017.

Braidotti, Rosi. *The Posthuman.* Cambridge: Polity Press, 2013.

Braidotti, Rosi. *Posthuman Knowledge.* Cambridge: Polity Press, 2019.

Braidotti, Rosi. "A Theoretical Framework for the Critical Posthumanities." *Theory, Culture and Society* 36, no. 6 (2018): 1–31.

Braidotti, Rosi. *Transpositions: On Nomadic Ethics.* Cambridge: Polity Press, 2008.

Braidotti, Rosi, and Simone Bignall. "Posthuman Systems." In *Posthuman Ecologies: Complexity and Process after Deleuze*, edited by Rosi Braidotti and Simone Bignall, 1–16. New York: Rowman and Littlefield International, 2019.

Butler, Judith. *Frames of War: When Is Life Grievable?* London: Verso, 2009.

Butler, Judith. "Hannah Arendt's Death Sentences." *Comparative Literature Studies* 48, no. 3 (2011): 280–95.

Bynkershoek, Cornelius van. *Cuius Est Solum Eius Est Usque Ad Coelum (Et Ad Inferos). A Photographic Reproduction of the Second Edition (1744).* Translated by Ralph Van Deman Magoffin. Edited by James Brown Scott. New York: Oxford University Press, 1923.

Carolan, Michael S. "From Patent Law to Regulation: The Ontological Gerrymandering of Biotechnology." *Environmental Politics* 17, no. 5 (2008): 749–65.

Casson, Louisa. *Greenpeace: Protect the Oceans. In Deep Water: The Emerging Threat of Deep-Sea Mining.* Greenpeace International, 2019.

Certeau, Michel de. *The Practice of Everyday Life.* Translated by Steven Rendall. Berkeley: University of California Press, 1984.

Cockell, Charles. "Extraterrestrial Liberty: Can It Be Planned?" In *Human Governance Beyond Earth: Implications for Freedom*, edited by Charles Cockell. Cham: Springer, 2015.

Coetzee, John M. *The Lives of Animals.* Princeton, NJ: Princeton University Press, 1999.

Collins, David A. "Efficient Allocation of Real Property Rights on the Planet Mars. Boston University." *Journal of Science and Technology Law* 14, no. 2 (2008): 201–19.

Collis, Christy. "Territories Beyond Possession? Antarctica and Outer Space." *The Polar Journal* 7, no. 2 (2017): 287–302.

Coombe, Rosemary J. "The Properties of Culture and the Politics of Possessing Identity: Native Claims in the Cultural Appropriation Controversy." *The Canadian Journal of Law and Jurisprudence* 6, no. 2 (1993): 249–85.

Crist, Eileen. "On the Poverty of Our Nomenclature." In *Anthropocene or Capitalocene? Nature, History, and the Crisis of Capitalism*, edited by Jason W. Moore. Oakland, CA: PM Press/Kairos, 2016.

Crutzen, Paul J. "Geology of Mankind: The Anthropocene." *Nature* 415, no. 6867 (2002): 23.

Cullinan, Cormac. *Wild Law: A Manifesto for Earth Justice*. Vermont: Chelsea Green, 2011.

Davenport, Christian. *The Space Barons: Jeff Bezos, Elon Musk, and the Quest to Colonize the Cosmos*. New York: Public Affairs, 2018.

Davies, Margaret. *Law Unlimited: Materialism, Pluralism, and Legal Theory*. Abingdon, UK: Routledge, 2017.

Davies, Margaret. *Property: Meanings, Histories, Theories*. Abingdon, UK: Routledge Cavendish, 2007.

Dawkins, Richard. *The Selfish Gene*. New York: Oxford University Press, 1989.

Deep Sea Mining Campaign, London Mining Network, and Mining Watch Canada. *Why the Rush? Seabed Mining in the Pacific Ocean*. 2019.

Delaney, David. "Afterword: Lively Ever After: Beyond the Cult of Immateriality." In *Animals, Biopolitics, Law: Lively Legalities*, edited by Irus Braverman. New York: Routledge, 2016.

Delaney, David. "Beyond the World: Law as a Thing of This World." In *Law and Geography*, edited by Carolyn Harrison and Jane Holder. Oxford: Oxford University Press, 2006.

Deleuze, Gilles. *Difference and Repetition*. Translated by Paul Patton. New York: Columbia University Press, 1994.

Deleuze, Gilles. *Foucault*. Translated by Seàn Hand. Minneapolis: University of Minnesota Press, 2006.

Deleuze, Gilles. "Immanence: A Life." In *Pure Immanence: Essays on a Life*. Translated by Anne Boyman. New York: Zone Books, 2012.

Deleuze, Gilles. "Jean Hyppolite's *Logic and Existence*." In *Desert Islands and Other Texts (1953–1974)*. Edited by David Lapoujade. Translated by Mike Taormina. Los Angeles: Semiotext(e), 2004.

Deleuze, Gilles. *The Logic of Sense*. Translated by Mark Lester and Charles Stivale. New York: Columbia University Press, 1990.

Deleuze, Gilles. *Nietzsche and Philosophy*. Translated by Hugh Tomlinson. New York: Columbia University Press, 2006.

Deleuze, Gilles. "Nomadic Thought." In *Desert Islands and Other Texts (1953–1974)*. Edited by David Lapoujade. Translated by Mike Taormina. Los Angeles: Semiotext(e), 2004.

Deleuze, Gilles. *Pure Immanence: Essays on a Life*. Translated by Anne Boyman. New York: Zone Books, 2012.

Deleuze, Gilles. *Spinoza: Practical Philosophy*. San Francisco: City Lights Books, 1988.

Deleuze, Gilles. "A Sunflower Seed Lost in a Wall Is Capable of Shattering That Wall." In *The Funambulist Pamphlets: Volume 3: Deleuze*, edited by Léopold Lambert. New York: Punctum Books, 2013. https://thefunambulistdotnet.wordpress.com/2013/07/05/funambulist-pamphlets-volume-03-deleuze-now-published/.

Deleuze, Gilles, and Michel Foucault. "Intellectuals and Power." In *Desert Islands and Other Texts (1953–1974)*, edited by David Lapoujade; translated by Mike Taormina. Los Angeles: Semiotext(e), 2004.

Deleuze, Gilles, and Félix Guattari. *Kafka: Toward a Minor Literature.* Translated by Dana Polan. Minneapolis: University of Minnesota Press, 1986.

Deleuze, Gilles, and Félix Guattari. *A Thousand Plateaus: Capitalism and Schizophrenia.* Translated by Brian Massumi. Minneapolis: University of Minnesota Press, 2014.

Deleuze, Gilles, and Félix Guattari. *What Is Philosophy?* Translated by Janis Tomlinson and Graham Burchell III. New York: Columbia University Press, 1994.

Deleuze, Gilles, and Claire Parnet. *Dialogues II.* New York: Columbia University Press, 2007.

de Man, Paul. "Autobiography as De-Facement." MLN 94, no. 5 (1979): 919–30.

Derman, Joshua. "Carl Schmitt on Land and Sea." *History of European Ideas* 37, no. 2 (2011): 181–9.

Derrida, Jacques. *The Animal That Therefore I Am.* Translated by Marie-Louise Mallet. New York: Fordham University Press, 2010.

Derrida, Jacques. "Eating Well." In *Who Comes After the Subject?*, edited by Eduardo Cadava, Peter Connor, and Jean-Luc Nancy. New York: Routledge, 1991.

Derrida, Jacques. "Following Theory." In *Life. After. Theory*, edited by Michael Payne and John Schad. London: Continuum, 2004.

Derrida, Jacques. "Force of Law: The Mystical Foundation of Authority." In *Deconstruction and the Possibility of Justice*, edited by Drucilla Cornell, Michel Rosenfeld, and David Gray Carlson. London: Routledge, 1992.

Derrida, Jacques. "The Law of Genre." *Critical Inquiry Critical Inquiry* 7, no. 1 (1980): 55–81.

Derrida, Jacques. *The Monolingualism of the Other: The Prosthesis of Origin.* Stanford, CA: Stanford University Press, 1998.

Derrida, Jacques. "No Apocalypse, Not Now (Full Speed Ahead, Seven Missiles, Seven Missives)." *Diacritics* 14, no. 2 (1984): 20–31.

Derrida, Jacques. "White Mythology: Metaphor in the Text of Philosophy." *New Literary History* 6, no. 1 (1974): 5–74.

Drake, Nadia. "Where, Exactly, Is the Edge of Space? It Depends on Who You Ask." *National Geographic*, December 20, 2018.

Dziak, Robert P., Joseph H. Haxel, Haruyoshi Matsumoto et al. "Ambient Sound at Challenger Deep, Mariana Trench." *Oceanography* 30, no. 2 (2017): 186–97.

Ehrenfried, Manfred von. *Stratonauts: Pioneers Who Ventured into the Stratosphere.* Cham: Springer International, 2014.

Erika, Lennon. "A Tale of Two Poles: A Comparative Look at the Legal Regimes in the Arctic and the Antarctic." *Sustainable Development Law and Policy* 8, no. 3 (2010): 32–36.

Esposito, Roberto. "The Dispositif of the Person." *Law, Culture and the Humanities* 8, no. 1 (2012): 17–30.

Esposito, Roberto. *Persons and Things: From the Body's Point of View.* John Wiley and Sons, 2015.

Esposito, Roberto. *Third Person: Politics of Life and Philosophy of the Impersonal.* Translated by Zakiya Hanafi. Cambridge: Polity, 2015.

Fanon, Frantz. *Black Skin, White Masks.* New York: Grove Atlantic, 2007.

Fanon, Frantz. *A Dying Colonialism.* New York: Grove Press, 1994.

Feynman, Richard P., Robert B. Leighton, and Matthew L. Sands. *The Feynman Lectures on Physics: New Millenium Edition*. Vol. 1, *Mainly Mechanics, Radiation, and Heat*. New York: Basic Books, 2010.

Fleurke, Floor, Michael Leach, Hans Lindahl, Phillip Paiement, Marie Petersmann, and Han Somsen. "Constitutionalizing in the Anthropocene." *Journal of Human Rights and the Environment* 15, no. 1 (February 29, 2024): 4–22.

Foucault, Michel. *The Order of Things: An Archaeology of the Human Sciences*. London: Routledge, 2006.

Foucault, Michel. Preface to *Anti-Oedipus: Capitalism and Schizophrenia*, by Gilles Deleuze and Félix Guattari. Translated by Brian Massumi. Minneapolis: University of Minnesota Press, 1983.

Foucault, Michel. "So Is It Important to Think?" Translated by James D. Faubion. In *Power*, edited by Paul Rabinow. Vol. 3 of *Essential Works of Foucault, 1954–1984*. London: Penguin, 2002.

Foucault, Michel. "What Is Critique?" In *What Is Enlightenment?: Eighteenth-Century Answers and Twentieth-Century Questions*, edited by James Schmidt. Berkeley: University of California Press, 1996.

Fountain, Lynn M. "Creating Momentum in Space: Ending the Paralysis Produced by the 'Common Heritage of Mankind" Doctrine.'" *Connecticut Law Review* 35, no. 4 (2003): 1753–87.

Fox, Warwick. "Foundations of a General Ethics: Selves, Sentient Beings, and Other Responsively Cohesive Structures." *Royal Institute of Philosophy Supplement* 69 (2011): 47–66.

Frisch, Max, and Heribert Kuhn. *Biedermann Und Die Brandstifter Ein Lehrstück Ohne Lehre*. Frankfurt am Main: Suhrkamp, 2014.

Frommherz, Gudrun. "Memetics of Transhumanist Imagery." *Visual Anthropology* 26, no. 2 (2013): 147–64.

Fuller, Steve, and Veronika Lipińska. *The Proactionary Imperative: A Foundation for Transhumanism*. Basingtoke: Palgrave Macmillan, 2014.

Galdamez, Laura. "Exploring the Stratosphere: What We Missed by Shooting for the Moon." In *Into Space: A Journey of How Humans Adapt and Live in Microgravity*, edited by Thais Russomano and Lucas Rehnberg. London: IntechOpen, 2018.

Ganchrow, Raviv. "Chlorophyll-Ocean-Iron-Breath." June 2022. https://ravivganchrow .com/page/iron/chl-ocn-irn-brt.html.

Ganchrow, Raviv. "Earth-Bound Sound: Oscillations of Hearing, Ocean, and Air." *Theory and Event* 24, no. 1 (2021): 67–116.

Gandorfer, Daniela. "Breathing Law: Real Imaginings of What It Might Mean to Matter Differently." In *The Cabinet of Imaginary Laws*, edited by Peter Goodrich and Thanos Zartaloudis. London: Routledge, 2021.

Gandorfer, Daniela. "Down and Dirty in the Field of Play: Startup Societies, Crypto-statecraft, and Critical Complicity." *Law and Critique* 33, no. 3 (2022): 355–77.

Gandorfer, Daniela. "Embodied Critique and Posthuman(ist) Legal Futures/Things Worth More Than Being Thought/Judge Schreber/Madness/Think!" In *Laws of Transgression: The Return of Judge Schreber*, edited by Peter Goodrich and Katrin Trüstedt. Toronto: University of Toronto Press, 2022.

Gandorfer, Daniela. "Introduction: What Is Your Power?" In *Research Handbook on Law and Literature*, edited by Peter Goodrich, Daniela Gandorfer and Cecilia Gebruers. Cheltenham, UK: Edward Elgar, 2022.

Gandorfer, Daniela, and Zulaikha Ayub. "Introduction: Matterphorical." In "Matterphorical," edited by Anker Libby and Cristina Beltrán. Special issue, *Theory and Event* 24, no. 1 (2021): 2–13.

Garforth, Kathryn. "Life as Chemistry or Life as Biology?: An Ethic of Patents on Genetically Modified Organisms." In *Patenting Lives: Life Patents, Culture and Development*, edited by Johanna Gibson. London: Routledge, 2008.

Geglia, Beth. "Honduras: Reinventing the Enclave." *NACLA Report on the Americas* 48, no. 4 (2016): 353–60.

Glissant, Édouard. *Poetics of Relation*. Translated by Betsy Wing. Ann Arbor: University of Michigan Press, 2010.

Goethe, Johann Wolfgang von. *Briefwechsel zwischen Schiller und Goethe in den Jahren 1794 bis 1805*. Edited by Karl Richter. Vol. 8.1 of *Sämtliche Werke Nach Epochen Seines Schaffens*. München: Hanser, 1990.

Goethe, Johann Wolfgang von. "Der Zauberlehrling." In *Goethe. Poetische Werke (Berliner Ausgabe)*. Berlin: Aufbau-Verlag, 1965.

Goethe, Johann Wolfgang von. "The Sorcerer's Apprentice." In *Goethe, the Lyrist: 100 Poems in New Translations Facing the Originals with a Biographical Introduction*, edited by Edwin Hermann Zeydel. Chapel Hill: University of North Carolina Press, 1965.

Goodman, Theodore W. "To the End of the Earth: A Study of the Boundary Between Earth and Space." *Journal of Space Law* 36, no. 1 (2010): 87–114.

Goodrich, Peter. *Advanced Introduction to Law and Literature*. Cheltenham, UK: Edward Elgar, 2021.

Goodrich, Peter. "The Critic's Love of the Law: Intimate Observations on an Insular Jurisdiction." *Law and Critique* 10, no. 3 (1999): 343–60.

Goodrich, Peter. "Critique of Comparative Law: To Compierre Negative Comparative Law: A Strong Programme for Weak Thought by Pierre Legrand." *Journal of Law and Society* 51, no. 1 (2024): 130–42.

Goodrich, Peter. "Transhumusians: On the Jurisography of the Corpus Iuris." *Theory and Event* 24, no. 1 (2021): 117–30.

Goodrich, Peter, Daniela Gandorfer, and Cecilia Gebruers, eds. *Research Handbook on Law and Literature*. Cheltenham, UK: Edward Elgar, 2022.

Grear, Anna. "Deconstructing Anthropos: A Critical Legal Reflection on Anthropocentric Law and Anthropocene Humanity." *Law and Critique* 26, no. 3 (2015): 225–49.

Grear, Anna. *Redirecting Human Rights: Facing the Challenge of Corporate Legal Humanity*. Basingstoke: Palgrave Macmillan, 2010.

Grear, Anna. "Towards 'Climate Justice'? A Critical Reflection on Legal Subjectivity and Climate Injustice: Warning Signals, Patterned Hierarchies, Directions for Future Law and Policy." In *Choosing a Future: The Social and Legal Aspects of Climate Change*, edited by Anna Grear and Conor Gearty. Cheltenham, UK: Edward Elgar, 2014.

Grosz, Elizabeth A. *The Incorporeal: Ontology, Ethics, and the Limits of Materialism*. New York: Columbia University Press, 2018.

Guattari, Félix. *The Anti-Œdipus Papers*. Translated by Kélina Gotman. New York: Semiotext(e), 2006.

Guerster, Markus, and Ulrich Walter. "Aerodynamics of a Highly Irregular Body at Transonic Speeds—Analysis of Stratos Flight Data." *PLOS One* 12, no. 12 (2017).

Halewood, Peter. "Law's Bodies: Disembodiment and the Structure of Liberal Property Rights." *Iowa Law Review* 81, no. 5 (1996).

Haraway, Donna J. *The Companion Species Manifesto: Dogs, People, and Significant Otherness*. Chicago: Prickly Paradigm Press, 2003.

Haraway, Donna J. *How Like a Leaf: An Interview with Thyrza Nichols Goodeve*. New York: Routledge, 2000.

Haraway, Donna J. *Modest_Witness@Second_Millennium: Femaleman_Meets_Oncomouse: Feminism and Technoscience*. London: Routledge, 1997.

Haraway, Donna J. "Situated Knowledges: The Science Question in Feminism and the Privilege of Partial Perspective." *Feminist Studies* 14, no. 3 (1988): 575–99.

Haraway, Donna J. *Staying with the Trouble: Making Kin in the Chthulucene*. Durham, NC: Duke University Press, 2016.

Haraway, Donna J., and Angela Creager. "Compost-Ography + Embodied Narratives." Introduction by Eduardo Cadava. Virtual lecture, December 2018, posted March 19, 2019, by Reading Matters, YouTube. https://www.youtube.com/watch?v=rwBANeOgEE4.

Hardack, Richard. "New and Improved: The Zero-Sum Game of Corporate Personhood." *Biography* 37, no. 1 (2014): 36–68.

Harney, Stefano, and Fred Moten. *The Undercommons: Fugitive Planning and Black Study*. Wivenhoe: Minor Compositions, 2013.

Harris, John. "Human Beings, Persons and Conjoined Twins: An Ethical Analysis of the Judgement in Re A." *Medical Law Review* 9, no. 3 (Autumn 2001): 221–36.

Harris, John. "Immortal Ethics." *Annals of the New York Academy of Sciences* 1019, no. 1 (2004): 527–34.

Hartman, Saidiya V., and Frank B. Wilderson. "The Position of the Unthought." *Qui Parle* 13, no. 2 (2003): 185.

Havice, Elizabeth, and Anna Zalik. "Ocean Frontiers: Epistemologies, Jurisdictions, Commodifications." *International Social Science Journal* 68, nos. 229–230 (2018): 219–35.

Hayles, N. Katherine. *How We Became Posthuman: Virtual Bodies in Cybernetics, Literature, and Informatics*. Chicago: University of Chicago Press, 1999.

Heffernan, Olive. "Deep-Sea Dilemma." *Nature* 571 (2019): 465–68.

Helmreich, Stefan. "Seagoing Nightmares." *Dialogues in Human Geography* 9, no. 3 (2019): 308–11.

Helmreich, Stefan. *Sounding the Limits of Life: Essays in the Anthropology of Biology and Beyond*. Princeton, NJ: Princeton University Press, 2016.

Helmreich, Stefan. "Wave Theory Social Theory." *Public Culture: Bulletin of the Project for Transnational Cultural Studies* 32, no. 2 (2020): 287–326.

Helmreich, Stefan, Daniela Gandorfer, and Zulaikha Ayub. "Doing Theory: Life, Ethics, and Force." *Theory and Event* 24, no. 1 (2021): 158–91.

Herbrechter, Stefan. "Narrating(—)Life—in Lieu of an Introduction." In *Narrating Life: Experiments with Human and Animal Bodies in Literature, Science and Art*, edited by Stefan Herbrechter and Elisabeth Friis. Leiden: Koninklijke Brill, 2016.

Herron, Thomas J. "Deep Space Thinking: What Elon Musk's Idea to Nuke Mars Teaches Us About Regulating the 'Visionaries and Daredevils' of Outer Space." *Columbia Journal of Environmental Law* 41, no. 3 (2019).

Hijiya, James A. "The 'Gita' of J. Robert Oppenheimer." *Proceedings of the American Philosophical Society* 144, no. 2 (2000): 123–67.

Holman, Christopher M. "Copyright for Engineered DNA: An Idea Whose Time Has Come?" *West Virginia Law Review* 699 (2010): 699–738.

Huff, Cynthia. "After Auto, After Bio: Posthumanism and Life Writing." *Auto/Biography Studies* 32, no. 2 (2017): 279–82.

Hund, Andrew J. "Antarctic, Definitions Of." In *Antarctica and the Arctic Circle*, edited by Andrew J. Hund. Santa Barbara: ABC-CLIO, 2014.

Hund, Andrew J. "Antarctic Territorial Claims." In *Antarctica and the Arctic Circle*, edited by Andrew J. Hund. Santa Barbara: ABC-CLIO, 2014.

Husserl, Edmund, and Karl Schuhmann. *Ideen Zu Einer Reinen Phänomenologie Und Phänomenologischen Philosophie. Erstes Buch, 1. Halbband*. The Hague: M. Nijhoff, 1976.

Impey, Chris. "Mars and Beyond: The Feasibility of Living in the Solar System." In *The Human Factor in a Mission to Mars: An Interdisciplinary Approach*, edited by Konrad Szocik. Cham: Springer, 2019.

Irigaray, Luce. *The Forgetting of Air in Martin Heidegger*. Translated by Mary Beth Mader. London: Athlone, 1999.

Irigaray, Luce, and Michael Marder. *Through Vegetal Being: Two Philosophical Perspectives*. New York: Columbia University Press, 2016.

Istel, Jacques-André. "Pour Le Sport." *Flying*, 1989, 28–29.

Jakosky, Bruce M., and Christopher S. Edwards. "Inventory of $CO_2$ Available for Terraforming Mars." *Nature Astronomy* 2, no. 8 (2018): 634–39.

James, Tom. *Deep Space Commodities: Exploration, Production and Trading*. London: Palgrave Macmillan, 2018.

Jasanoff, Sheila. *Can Science Make Sense of Life?* Cambridge: Polity Press, 2019.

Jasanoff, Sheila. "Conclusion: Rewriting Life, Reframing Rights." In *Reframing Rights: Bioconstitutionalism in the Genetic Age*, edited by Sheila Jasanoff. Cambridge, MA: MIT Press, 2011.

Jasanoff, Sheila. "Taking Life: Private Rights in Public Nature." In *Lively Capital: Biotechnologies, Ethics, and Governance in Global Markets*, edited by Sunder Rajan Kaushik. Durham, NC: Duke University Press, 2012.

Jaszi, Peter, and Martha Woodmansee. *The Construction of Authorship: Textual Appropriation in Law and Literature*. Durham, NC: Duke University Press, 1999.

Jhering, Rudolf von. "Im Juristischen Begriffshimmel. Ein Phantasiebild." In *Scherz Und Ernst in Der Jurisprudenz. Eine Weihnachtsgabe Für Das Juristische Publikum Von Rudolf Von Ihering*, 244–334. Leipzig: Druck und Verlag von Breitkopf & Härtel, 1884.

Johns, Fleur. "From Planning to Prototypes: New Ways of Seeing Like a State." *Modern Law Review* 82, no. 5 (2019): 833–63.

Joiner, James T., and NOAA Diving Program. *NOAA Diving Manual*. Flagstaff, AZ: Best, 2001.

Jones, David S., and Stefan Helmreich. "The Shape of Epidemics." *Boston Review: A Political and Literary Forum*, June 26, 2020.

Jones, Owain. "Lunar–Solar Rhythmpatterns: Towards the Material Cultures of Tides." *Environment and Planning A: Economy and Space* 43, no. 10 (2011): 2285–303.

Jones, Reilly. "A Critique of the Declaration of the Independence of Cyberspace." *Extropy: The Journal of Transhumanist Thought*, Winter 1996, 28–31.

Jones, Reilly. "A History of Extropic Thought: Parallel Conceptual Development of Technicism and Humanism." Paper presented at the Extro^2 Conference, Santa Monica, CA, June 18, 1995. https://www.reillyjones.com/history-of-extropic -thought.html.

Joyner, Christopher C. "Ice-Covered Regions in International Law." *Natural Resources Journal* 31, no. 1 (1991): 213–42.

Kakaes, Konstantin. "Five Schemes for Cheaper Space Launches—and Five Cautionary Tales." *MIT Technology Review*, June 26, 2019.

Keller, Catherine. *Cloud of the Impossible. Negative Theology and Planetary Entaglement*. New York: Columbia University Press, 2017.

Keller, Evelyn Fox. *Making Sense of Life: Explaining Biological Development with Models, Metaphors and Machines*. Cambridge, MA: Harvard University Press, 2003.

Klein, Natalie. *Maritime Security and the Law of the Sea*. New York: Oxford University Press, 2011.

Kneubuehl, Beat P. *Wound Ballistics: Basics and Applications*. Berlin: Springer, 2011.

Kniffin, Kevin M., Brian Wansink, Vladas Griskevicius, and David Sloan Wilson. "Beauty Is in the In-Group of the Beholded: Intergroup Differences in the Perceived Attractiveness of Leaders." *The Leadership Quarterly* 25, no. 6: 1143–53.

Koskenniemi, Martti. *From Apology to Utopia: The Structure of International Legal Argument*. Cambridge: Cambridge University Press, 2015.

Krebs, Robert E. *The History and Use of Our Earth's Chemical Elements: A Reference Guide*. Westport, CT: Greenwood Press, 2006.

Kuokkanen, Rauna Johanna. *Reshaping the University: Responsibility, Indigenous Epistemes, and the Logic of the Gift*. Vancouver: UBC Press, 2007.

Lal, Bhavya, and Emily Nightingale. "Where Is Space? And Why Does That Matter." Paper presented at Space Traffic Management Conference: Roadmap to the Stars, Embry–Riddle Aeronautical University, Daytona Beach, FL, 2014.

Latour, Bruno. *Facing Gaia: Eight Lectures on the New Climate Regime*. Translated by Catherine Porter. Cambridge: Polity Press, 2017.

Lehman, Jessica, Philip Steinberg, and Elizabeth R. Johnson. "Turbulent Waters in Three Parts." *Theory and Event* 24, no. 1 (2021): 192–219.

Lejeune, Philippe, and Paul John Eakin. *On Autobiography*. Minneapolis: University of Minnesota Press, 1989.

Lewis, Cory J. "Iceberg Harvesting: Suggesting a Federal Regulatory Regime for a New Freshwater Source." *Boston College Environmental Affairs Law Review* 42, no. 2 (2015): 439–72.

Li, Maggie, Markus Hilpert, Jeff Goldsmith et al. "Air Pollution in American Indian Versus Non-American Indian Communities, 2000–2018." *American Journal of Public Health* 112, no. 4 (2022): 615–23.

Lukin, Valery. "The Antarctic Treaty System (ATS)." In *Antarctica and the Arctic Circle*, edited by Andrew J. Hund. Santa Barbara: ABC-CLIO, 2014.

Lyall, Francis, and Paul B. Larsen. *Space Law: A Treatise*. Farnham: Ashgate, 2009.

MacDonald, Fraser. "Anti-Astropolitik—Outer Space and the Orbit of Geography." *Progress in Human Geography* 31, no. 5 (2007): 592–615.

Manning, Erin. "Another Regard." In *The Lure of Whitehead*, edited by Nicholas Gaskill and A. J. Nocek. Minneapolis: University of Minnesota Press, 2014.

Manning, Erin. *Politics of Touch: Sense, Movement, Sovereignty*. Minneapolis: University of Minnesota Press, 2007.

Mansfield, Becky. "Neoliberalism in the Oceans: 'Rationalization,' Property Rights, and the Commons Question." *Geoforum* 35, no. 3 (2004): 313–26.

Massumi, Brian. *99 Theses on the Revaluation of Value: A Postcapitalist Manifesto*. Minneapolis: University of Minnesota Press, 2018.

Massumi, Brian. *The Principle of Unrest: Activist Philosophy in the Expanded Field*. London: Open Humanities Press, 2017.

Mawani, Renisa. *Across Oceans of Law: The Komagata Maru and Jurisdiction in the Time of Empire*. Durham, NC: Duke University Press, 2018.

Maxwell, Lynn M. *Wax Impressions, Figures, and Forms in Early Modern Literature: Wax Works*. Cham: Palgrave Macmillan, 2020.

Mbembe, Achille. *Critique of Black Reason*. Translated by Laurent Dubois. Johannesburg: Wits University Press, 2017.

Mbembe, Achille. "Decolonizing the University: New Directions." *Arts and Humanities in Higher Education* 15, no. 1 (2016): 29–45.

Metzler, Ingrid. "Human Life between Biology and Law in Germany." In *The Routledge Handbook of Biopolitics*, edited by Sergei Prozorov and Simona Rentea. London: Routledge, 2017.

Metzler, Ingrid, and Andrew Webster. "Bio-Objects and Their Boundaries: Governing Matters at the Intersection of Society, Politics, and Science." *Croatian Medical Journal* 52, no. 5 (2011): 648–50.

Mitchell, Timothy. *Carbon Democracy: Political Power in the Age of Oil*. London: Verso, 2013.

Moore, Jason W. "The Rise of Cheap Nature." In *Anthropocene or Capitalocene?: Nature, History, and the Crisis of Capitalism*, edited by Jason W. Moore. Oakland: PM Press/Kairos, 2016.

Moore-Gilbert, B. J. *Postcolonial Life-Writing: Culture, Politics, and Self Representation*. New York: Routledge, 2009.

Moravec, Hans. *Mind Children: The Future of Robot and Human Intelligence*. Cambridge, MA: Harvard University Press, 1988.

More, Max. "The Extropian Principles." *Extropy: Vaccine for Future Shock*, Summer 1990, 17–18.

More, Max. "Extropian Principles 2.5." *Extropy: The Journal of Transhumanist Thought*, Summer/Fall 1993, 9–12.

More, Max. "The Heat Death of Timothy Leary." *Extropy: The Journal for Transhumanist Thought*, Winter 1996, 37–39.

More, Max. "A Letter to Mother Nature." In *The Transhumanist Reader: Classical and Contemporary Essays on the Science, Technology, and Philosophy of the Human Future*, edited by Max More and Natasha Vita-More. Chichester, West Sussex: Wiley-Blackwell, 2013.

More, Max. "Order Without Orders." *Extropy: The Journal for Transhumanist Thought*, Spring 1991, 21–32.

More, Max. "The Philosophy of Transhumanism." In *The Transhumanist Reader: Classical and Contemporary Essays on the Science, Technology, and Philosophy of the Human Future*, edited by Max More and Natasha Vita-More. Oxford: John Wiley and Sons, 2013.

More, Max. "Transhumanism: Towards a Futurist Philosophy." *Extropy: Vaccine for Future Shock*, Summer 1990, 6–12.

Moten, Fred. "Blackness and Nothingness (Mysticism in the Flesh)." *South Atlantic Quarterly* 112, no. 4 (2013): 737–80.

Naffine, Ngaire. "The Body Bag." In *Sexing the Subject of Law*, edited by Ngaire Naffine and Rosemary Owens. North Ryde: LBC Information Services and Sweet and Maxwell, 1997.

Naffine, Ngaire. *Law's Meaning of Life: Philosophy, Religion, Darwin and the Legal Person*. Oxford: Hart, 2009.

Naffine, Ngaire. "Legal Personality and the Natural World: On the Persistence of the Human Measure of Value." *Journal of Human Rights and the Environment* 3 (2012): 68–83.

Nagel, Thomas. "What Is It Like to Be a Bat?" *The Philosophical Review* 83, no. 4 (1974): 435–50.

Naimou, Angela. *Salvage Work: U.S. And Caribbean Literatures Amid the Debris of Legal Personhood*. New York: Fordham University Press, 2017.

National Lawyers Guild. *Report of the National Lawyers Guild Delegation Investigation of Zones for Economic Development and Employment in Honduras*. New York: National Lawyers Guild, 2014.

Negri, Antonio. *The Savage Anomaly: The Power of Spinoza's Metaphysics and Politics*. Translated by Michael Hardt. Minneapolis: University of Minnesota Press, 1991.

Nelkin, Dorothy. "Molecular Metaphors: The Gene in Popular Discourse." *Nature Reviews: Genetics* 2, no. 7 (2001): 555–59.

Neumann, Gerhard. "Schillers Dramatische Fragmente. Ein Projekt Der Moderne." In *Friedrich Schiller Und Der Weg in Die Moderne*, edited by Walter Hinderer. Würzburg: Königshausen et Neumann, 2006.

Nietzsche, Friedrich. "On Truth and Lie in an Extra-Moral Sense." Translated by Walter Kaufmann. In *The Portable Nietzsche*, edited by Walter Kaufmann. New York: Viking Press, 1982.

O'Brien, Zeldine. "Advertising in Space. Sales at the Outer Limits." In *Commercial Space Exploration. Ethics, Policy and Governance*, edited by Jai Galliott. Farnham: Ashgate, 2015.

O'Connell, Daniel Patrick. "The Extent of the Territorial Sea." In *The International Law of the Sea*, edited by I. A. Shearer. Oxford: Oxford University Press, 1982.

O'Connor, Max T. "Space Colonization." *Extropy: Vaccine for Future Shock*, Fall 1988, 11–12.

O'Connor, Max T. "What's Wrong with Death?" *Extropy: Vaccine for Future Shock*, Summer 1989, 20–28.

O'Connor, Max T., and Thomas W. Bell. "The Extropian Declaration." *Extropy: Vaccine for Future Shock*, Winter 1990, 51.

O'Connor, Max T., and Thomas W. Bell. "Introduction." *Extropy: Vaccine for Future Shock*, Fall 1988, 1–13.

O'Connor, Max T., and Thomas W. Bell. "Mindfucking." *Extropy: Vaccine for Future Shock*, Fall 1988, 10.

Ofrias, Lindsay, and Gordon Roecker. "Organized Criminals, Human Rights Defenders, and Oil Companies: Weaponization of the Rico Act across Jurisdictional Borders." *Focaal*, no. 85 (2019): 37–50.

O'Neill, Gerard K. *The High Frontier: Human Colonies in Space*. Burlington: Apogee, 2000.

Otto, Beate. *Unterwasser-Literatur: Von Wasserfrauen Und Wassermännern*. Würzburg: Königshausen und Neumann, 2001.

Persson, Erik. "Citizens of Mars Ltd." In *Human Governance Beyond Earth: Implications for Freedom*, edited by Charles Cockell, 121–37. Cham: Springer, 2015.

Peters, Kimberley. "Manipulating Material Hydro-Worlds: Rethinking Human and More-Than-Human Relationality Through Offshore Radio Piracy." *Environment and Planning A: Economy and Space* 44, no. 5 (2012): 1241–54.

Philippopoulos-Mihalopoulos, Andreas. "Critical Autopoiesis and the Materiality of Law." *International Journal for the Semiotics of Law* 27, no. 2 (2014): 389–418.

Philippopoulos-Mihalopoulos, Andreas. "Lively Agency: Life and Law in the Anthropocene." In *Animals, Biopolitics, Law: Lively Legalities*, edited by Irus Braverman. New York: Routledge, 2016.

Philippopoulos-Mihalopoulos, Andreas. *Spatial Justice: Body, Lawscape, Atmosphere*. London: Routledge, 2015.

Philippopoulos-Mihalopoulos, Andreas. "To Have to Do with the Law. An Essay." In *Routledge Handbook of Law and Theory*, edited by Andreas Philippopoulos-Mihalopoulos. New York: Routledge, 2019.

Pottage, Alain. "A Unique and Different Subject of Law." *Cardozo Law Review* 16, nos. 3–4 (1995): 1161–204.

Povinelli, Elizabeth A. *Between Gaia and Ground: Four Axioms of Existence and the Ancestral Catastrophe of Late Liberalism*. Durham, NC: Duke University Press, 2021.

Povinelli, Elizabeth A. *Geontologies: A Requiem to Late Liberalism*. Durham, NC: Duke University Press, 2016.

Povinelli, Elizabeth A. "On Biopolitics and the Anthropocene: Elizabeth Povinelli, interviewed by Kathryn Yusoff and Mat Coleman." *Society and Space*, March 7, 2014.

Povinelli, Elizabeth A., Daniela Gandorfer, and Zulaikha Ayub. "Mattering-Forth: Thinking-with Karrabing." *Theory and Event* 24, no. 1 (2021): 294–323.

Rasmussen, M.B., and C. Lund. "Reconfiguring Frontier Spaces: The Territorialization of Resource Control." *World Development* 101 (2018): 388–99.

Redfield, Peter. "The Half-Life of Empire in Outer Space." *Social Studies of Science* 32, nos. 5–6 (2002): 791–825.

Roosth, Sophia. "Life, Not Itself: Inanimacy and the Limits of Biology." *Grey Room*, no. 57 (2014): 56–81.

Roosth, Sophia. *Synthetic: How Life Got Made*. Chicago: University of Chicago Press, 2017.

Rosenfield, Stanley B. "Where Air Space Ends and Outer Space Begins." *Journal of Space Law* 7 (1979): 137–48.

Ross, Benjamin. *The Philosophy of Transhumanism: A Critical Analysis*. Bingley: Emerald, 2020.

Ryan, Craig. *Magnificent Failure: Free Fall from the Edge of Space*. Washington, DC: Smithsonian Books, 2003.

Ryan, Michael. *Politics and Culture: Working Hypotheses for a Post-Revolutionary Society*. Baltimore: Johns Hopkins University Press, 1989.

Schabas, William A. "Right to Life." In *Encyclopedia of Human Rights*, edited by David P. Forsythe. Oxford: Oxford University Press, 2009.

Schiller, Friedrich. "Die Verschwörung Des Fiesko Zu Genua." In *Werke Und Briefe: In 12 Bänden, Friedrich Schiller, Dramen I*, edited by Hitzinger et al. Frankfurt am Main: Deutscher Klassiker-Verlag, 1988.

Schiller, Friedrich von. "Der Taucher." In *Schillers Werke: Nationalausgabe. Erster Band, Gedichte 1776–1799*, edited by Julius Petersen and Friedrich Beißner. Weimar: Hermann Böhlaus Nachfolger, 1992.

Schmitt, Carl. *Carl Schmitt: Antworten in Nürnberg*. Edited by Helmut Quaritsch. Berlin: Duncker and Humblot, 2000.

Schmitt, Carl. *Der Nomos Der Erde Im Völkerrecht Des Jus Publicum Europaeum*. Berlin: Duncker and Humblot, 2011.

Schmitt, Carl. *Land and Sea: A World-Historical Meditation*. Candor: Telos Press, 2015.

Schmitt, Carl. *Land Und Meer: Eine Weltgeschichtliche Betrachtung*. Stuttgart: Klett-Cotta, 2018.

Schmitt, Carl. *The Nomos of the Earth in the International Law of the Jus Publicum Europaeum*. New York: Telos Press, 2006.

Scott, James C. *Seeing Like a State: How Certain Schemes to Improve the Human Condition Have Failed*. New Haven, CT: Yale University Press, 2020.

Seeley, Thomas D. *Honeybee Democracy*. Princeton, NJ: Princeton University Press, 2010.

Shiva, Vandana. *Biopiracy: The Plunder of Nature and Knowledge*. Berkeley, CA: North Atlantic Books, 2016.

Shiva, Vandana. *Who Really Feeds the World?: The Failures of Agribusiness and the Promise of Agroecology*. Berkeley, CA: North Atlantic Books, 2016.

Skelton, Randall R. *A Survey of the Forensic Sciences*. Raleigh, NC: Lulu, 2010.

Slaughter, Joseph R. *Human Rights, Inc.: The World Novel, Narrative Form, and International Law*. New York: Fordham University Press, 2009.

Smith, Sidonie, and Julia Watson. *De/Colonizing the Subject: The Politics of Gender in Women's Autobiography*. Minneapolis: University of Minnesota Press, 1992.

Smith, Sidonie, and Julia Watson. *Reading Autobiography: A Guide for Interpreting Life Narratives*. Minneapolis: University of Minnesota Press, 2010.

Soucek, Alexander. "The Polar Regions." In *Outer Space in Society, Politics and Law*, edited by Christian Brünner and Alexander Soucek. Vienna: Springer, 2011.

Sparrow, Robert. "A Not-So-New Eugenics." *Hastings Center Report* 41, no. 1 (2011): 32–42.

Spinoza, Benedictus de. *Ethics: Proved in Geometrical Order*. Cambridge: Cambridge University Press, 2018.

Srinivasan, Balaji. *The Network State*. 2022. https://thenetworkstate.com/book/tns.pdf.

Steinberg, Philip E. "Of Other Seas: Metaphors and Materialities in Maritime Regions." *Atlantic Studies* 10, no. 2 (2013): 156–69.

Steinberg, Philip E. "Sovereignty, Territory, and the Mapping of Mobility: A View from the Outside." *Annals of the Association of American Geographers* 99, no. 3 (2009): 467–95.

Steinberg, Philip E., and Kimberley Peters. "Wet Ontologies, Fluid Spaces: Giving Depth to Volume Through Oceanic Thinking." *Environment and Planning D: Society and Space* 33, no. 2 (2015): 247–64.

Sterns, P.M., and L.I. Tennen. "Privateering and Profiteering on the Moon and Other Celestial Bodies: Debunking the Myth of Property Rights in Space." *Advances in Space Research* 31, no. 11 (2003): 2433–40.

Stevens, Adam H. "The Price of Air." In *Human Governance Beyond Earth: Implications for Freedom*, edited by Charles Cockell. Cham: Springer, 2015.

Straight, Nathan. *Autobiography, Ecology, and the Well-Placed Self: The Growth of Natural Biography in Contemporary American Life Writing*. New York: Peter Lang, 2011.

Sublette, Carey, and Scientists Federation of American. "The High Energy Weapons Archive: A Guide to Nuclear Weapons." 2000. http://www.ciar.org/ttk/hew/hew/index.html.

Sullivan, Mark. "Zedes: Neocolonialism and Land Grabbing in Honduras." Documentary, produced by National Lawyers Guild, posted October 16, 2015, by Mark Sullivan. YouTube, https://www.youtube.com/watch?v=_kRRnBIbgzo.

TallBear, Kimberly. "Beyond the Life/Not Life Binary: A Feminist-Indigenous Reading of Cryopreservation, Interspecies Thinking and the New Materialisms." In *Cryopolitics, Frozen Life in a Melting World*, edited by Joanna Radin and Emma Kowal. Cambridge, MA: MIT Press, 2017.

Taylor, Cynthia, and Bryan M. Dewsbury. "On the Problem and Promise of Metaphor Use in Science and Science Communication." *Journal of Microbiology and Biology Education* 19, no. 1 (2018): 1–5.

Thaman, Konai Helu. "Thinking." In *Songs of Love: New and Selected Poems, 1974–1999*. Suva: Mana Publications, 1999.

Tsing, Anna Lowenhaupt. "Natural Resources and Capitalist Frontiers." *Economic and Political Weekly* 38, no. 48 (2003): 5100–5106.

Turner, Frederick Jackson. "The Significance of the Frontier in American History." In *The Frontier in American History*. New York: Holt, 1920.

Uexküll, Jakob von. *A Foray into the Worlds of Animals and Humans: With a Theory of Meaning*. Translated by Joseph D. O'Neill. Minneapolis: University of Minnesota Press, 2010.

United States Environmental Protection Agency. *Policy Assessment for the Review of the National Ambient Air Quality Standards for Particulate Matter*. EPA-452/R-20–002 (2020).

Valentine, David. "Exit Strategy: Profit, Cosmology, and the Future of Humans in Space." *Anthropological Quarterly* 85, no. 4 (2012): 1045–67.

Viikari, L.E. "The Legal Regime for Moon Resource Utilization and Comparable Solutions Adopted for Deep Seabed Activities." *Advances in Space Research* 31, no. 11 (2003): 2427–32.

Vitruvius, Pollio. *Vitruvius: Ten Books on Architecture*. Translated by Thomas Noble Howe. New York: Cambridge University Press, 2007.

Viveiros de Castro, Eduardo. *Cannibal Metaphysics: For a Post-Structural Anthropology*. Translated by Peter Skafish. Minneapolis: Univocal, 2015.

Waldron, Jeremy. "Thoughtfulness and the Rule of Law." *British Academy Review*, no. 18 (2011): 1–11.

Weheliye, Alexander G. *Habeas Viscus: Racializing Assemblages, Biopolitics, and Black Feminist Theories of the Human*. Durham, NC: Duke University Press, 2014.

Weil, Kari. *Autobiography: Literature and the Posthuman*. Edited by Bruce Clarke and Manuela Rossini. Cambridge: Cambridge University Press, 2017.

Weinzierl, Matthew. "Space, the Final Economic Frontier." *Journal of Economic Perspectives* 32, no. 2 (2018): 173–92.

Weizman, Eyal. *Forensic Architecture: Violence at the Threshold of Detectability*. New York: Zone Books, 2018.

Weizman, Eyal, and Fazal Sheikh. *The Conflict Shoreline: Colonization as Climate Change in the Negev Desert*. Göttingen: Steidl, 2015.

Whitehead, Alfred North. *The Function of Reason: Louis Clark Vanuxem Foundation Lectures, Delivered at Princeton University, March 1929*. Princeton, NJ: Princeton University Press, 1929.

Whitehead, Alfred North. *Modes of Thought*. New York: Free Press, 1968.

Whitehead, Alfred North. *Process and Reality*. New York: Free Press, 2010.

Whitehead, Alfred North. *Science and the Modern World*. New York: Free Press, 2008.

Whitlock, Gillian. *Postcolonial Life Narratives: Testimonial Transactions*. New York: Oxford University Press, 2015.

Wilcox, Len. *Desert Dancing*. Edison, NJ: Hunter, 2009.

Williams, Patricia J. *The Alchemy of Race and Rights*. Cambridge, MA: Harvard University Press, 1999.

Williams, Patricia J. "Skittles as Matterphor." *Theory and Event* 24, no. 1 (2021): 356–98.

Wilson, Shawn. "What Is an Indigenous Research Methodology?" *Canadian Journal of Native Education* 25, no. 2 (2001): 175–79.

Wolf, Christa. *Kein Ort. Nirgends*. Frankfurt am Main: Suhrkamp, 2007.

Wolf, Christa. *No Place on Earth*. Translated by Jan van Heurck. New York: Farrar, Straus and Giroux, 1982.

Wolfe, Cary. *What Is Posthumanism?* Minneapolis: University of Minnesota Press, 2010.

Zhang, Wanlu Laura. "Extraterritorial Jurisdiction on Celestial Bodies." *Space Policy* 47 (2019): 148–57.

# Index

autonomous underwater vehicles (AUVs), 161
autonomy, 196
Ayub, Zulaikha, 11

Babylonian Code of Hammurabi, 113
Ball, Philip, 175
Barad, Karen: agential realism, 114–15;
    animate or inanimate, 67; brittle star,
    36–37, 43, 45; differentiating, 55, 186;
    ethics of thought, 94; forces, 34, 100;
    immanent critique, 76; knowing, ontology
    of, 35; on language, 33; onto-epistemology,
    52; posthumanism, 39, 41; quantum-field,
    159; spacetimemattering, 150; subject-
    object distinction, 47, 48; on theory, 20,
    56; the void, 38, 52, 54, 66
Barlow, John Perry, 108, 119
Batteux, Charles, 99
Baumgartner, Felix, 8, 103, 107–17, 120,
    128, 133–36, 151
beginnings, 7–10
being, modes of, 9
being-thought, 64, 67, 69
Bell, Thomas W., 77, 80, 88–91, 201
Benjamin, Marina, 129
Benjamin, Walter, 100
Bergson, Henri, 39
Bezos, Jeff, 136
Bignall, Simone, 40
Bikini Atoll, 33
binaries: relations, 47; thought, 64
biocapitalism, 176; logic, 180
biocide, 158
biodiversity, 170, 181
biography, 197
biology, 175; determinism, 42, 46
biomedia, 177
*bio nullius*, 171
bioprospecting, 170
biosciences, 176
biosecurity, 170
biotechnology, 81, 170, 180
biotransfer, 170
Bitcoin, 139
Black people, 102; Americans, 208;
    community, 77; thought, 74
blockchain, 11, 81, 122, 124, 138–39;
    governance, 93

Blomqvist, Jan, 119
Bloshenko, Aleksandr, 104
Blue Origin, 136
Böhmermann, Jan, 113
borders, 9, 93, 117, 142
Bostrom, Nick, 87, 199–200
boundaries, 141, 151–53, 157, 192;
    drawing, 140, 162
Boyle's law, 4, 147
BP, 198
Braidotti, Rosi, 38, 40, 102, 196
brain-computer interfacing, 106
Branson, Richard, 136
Brazil, 133
breathing, 23, 42, 51, 133, 135, 148–49,
    208, 212–17
Brentano, Bettina, 71
Brentano, Clemens, 71
brittle star, 35–37, 43–51, 75, 174.
    *See also Ophiomastix wendtii*
Butler, Judith, 18, 72
Bynkershoek, Cornelius van, 144

Cameron, James, 160
Canada, 155
cannons, 145; cannon-shot rule, 144
capitalism, 39, 92, 101; extraction, 158;
    logic, 171; radical, 87
carbon, 145
carbon dioxide, 148, 214
Cardozo Law School, 31
Carson, Anne, 2
Cartesianism, 169–70, 189–92, 201; cut, 47,
    52, 180, 184, 204, 206; law, 154; mind-body
    dualism, 8, 35, 52, 80, 86, 135, 142, 199;
    mode of thinking, 20, 24, 32; truth, 65;
    word/world separation, 149
cartography, 152
Castro, Xiomara, 91
causality, 67
Celestis, 138
centralized institutions, 10
charter cities, 93
Chauvin, Derek, 212
chemoheterotrophs, 192
Chevron, 198
China, 106
Christianity, 116

Hegel, 65
Heidegger, Martin, 24, 30, 143, 149
Helium, 149, 214
Helmreich, Stefan, 152, 170, 177
Henry's law, 147
Herbrechter, Stefan, 193
high seas, 142, 162–63
Hiroshima, 33, 150
Hispanic, 208
Hodl the void, 125
Hofer, Norbert, 116
Holocaust, 18
Holocene, 169
Homer, 148
homesteading, 129
homogenization, 170, 186
homosphere, 157, 172, 192
hostile universe, 22
Huff, Cynthia, 196
human, 177, 180, 187, 190–91, 194, 197–99,
    207; enhancement, 139; exceptionalism, 39,
    169; knowledge, 31; more-than-, 38.
    See also posthumanism; speciesism
humanism, 39–40, 81, 87–88, 190;
    assumptions, 41; classical, 20;
    Enlightenment, 22; modes of
    thinking, 93; thought, 79, 108
humanities: classical, 58; scholars, 29–30
Humanities Sequence, 58
Humphrey, N. K., 84
Hund, Andrew J., 155
Husserl, Edmund, the tree, 34
hydroacoustics, 160–61
hydrogen, 149, 153, 214; bomb, 33
hydropower, 148
Hyppolite, Jean, 52

I, 187–88, 194
#icantbreathe, 208, 212, 216
Icarus, 100, 103
Iceland, 155
ice shelves, 156
Identitarian Movement Austria, 116
identity, 59
ignorance, 62
image of thought, 63–65, 69–71, 78–79, 101,
    183, 206, 214–15; classical, 181, 187;
    destruction of, 71–72, 75

immanence, 74, 79; plane of, 24, 38, 47,
    52–53, 64–66, 69, 75, 159, 174; thinkers, 172
immanent critique, 76
immortalists, 85, 178
immortality, 189
imperialism, 62, 128, 192; language, 44.
    See also language
Impey, Chris, 136
inclusion, 11, 192
indeterminacy, 140, 159–63, 203–9
indigenous people, 102, 177, 208, 215;
    thinkers, 197; thought, 38
individualism, 169, 191, 194, 196; possessive,
    181; radical, 86
individualist anarchism, 131
inequality, 29, 141, 169, 217
insane diagrams, 71
inseparability, 28
intellectual crime, 19
interconnectedness, 196
International Air Sports Federation, 141
international contracts, 10
International Convention for the Prevention
    of Pollution from Ships, 155
International Seabed Authority (ISA), 161
International Space Station (ISS), 120
IoT (Internet of Things), 11
Irigaray, Luce, 24, 149–50
irony, weapons of, 29, 32
*irrsinnige Pläne. See* insane diagrams
Islam, 116
Israel, 163
Istel, Jacques-André, 112–13

Jasanoff, Sheila, 171, 176
Jasz, Peter, 181
J. Craig Venter Institute (JCVI), 181
Jemisin, N. K., 16
Jhering, Rudolf von, 100
Johnson, Elizabeth R., 152
Jones, Owein, 145
Jones, Reilly, 132
Joyner, Christopher C., 156
judgmental reasoning, 139
jurisdiction, 141; political, 131
jurisprudence, 157, 171; medical, 189
justice, 27, 51, 197, 206; decentralized, 11;
    representational, 32

NewSpace, 129
Newtonianism, 115
NFTS, 125
Nietzsche, Friedrich, 27, 31, 41, 65, 82
nitrogen, 149, 214
nitrous oxide, 149
Nitsch, Herbert, 144
N-methylformamide, 139
Nolan, Chris, 89
nomad thought, 65. *See also* counterthought
Nomocene, 172
nonappropriation principle, 130
nonbinary relationality, 79
nonexclusive knowledge production, 102
nonhuman, 38–39, 94, 136–37, 140; alliance
    with, 207; cryopreservation, 177; empathy,
    85; forces, 31–32; law, 150, 172, 184, 180;
    materialization of bodies, 114; molecular
    level, 190–91, 194–99; posthumanism, 41;
    position of power, 70; sentience, 187. *See
    also* posthumanism
nonhumanism, 40
nonlife, 184–90, 206; autobiography, 193,
    209; law, 150, 173; separation from life,
    170–72, 176–80, 196–97, 201
nonrepresentationality, 35; mode of thought,
    17, 32, 59, 63–76; relationality, 45
normativity, 25, 159; frames, 180; mode of, 158
Norris, Christopher, 42
North Pole, 155
Norway, 155
novelty, 61
nuclear: criticism, 29–30; equation, 30;
    fission, 28, 33; fusion, 33; metaphor of, 33;
    weapons, 33
Nuremberg, 19

Obama, Barack, 113
obedience, 146
observed constants, 131
Ochinero, Tom, 123
O'Connor, Max T., 77, 80. *See also*
    More, Max
OncoMouse, 181
O'Neill, Gerard K., 129, 138
onto-epistemology, 47, 49, 52, 62, 65, 75, 94,
    158, 179–80, 204; assumptions, 108, 132,
    142, 157, 159; cuts, 9, 191, 197, 201, 206;

difference, 61, 68, 102; expression, 148;
    expression of existence, 159; normativity,
    29; relationality, 39, 41, 50, 154
ontology, 16, 45, 52, 70; assumptions, 135, 159;
    entanglements, 47; killing, 152; thought,
    67; wet, 152. *See also* onto-epistemology
*Ophiomastix wendtii*, 35, 50
Oppenheimer, J. Robert, 33
oppression, 21, 75, 163
Orbán, Viktor, 116
Orbital Display, 121
origins, 8
Other, the, 28, 74
outer space, 120–24, 128–30, 134, 138,
    140–42, 149, 153, 162–63, 190;
    colonization, 137
Outer Space Treaty (OST), 121, 144
oxygen, 137–38, 145, 147–49, 153, 214

Pacific Ocean, 162
Palestine, property rights, 163
Palmer, Jackson, 122
Pantaleo, Daniel, 143, 208
Papua New Guinea, 162
Parnet, Claire, 17, 183
patriarchal patterns, 24
persistence, art of, 60
personhood, 191, 193, 198–99, 215
perspectivism, 43–44
Peters, Kimberley, 152
PG&E, 198
Philippopoulos-Mihalopoulos,
    Andreas, 28, 159, 172, 204
philology, 6
philosophy, 7; thought, 149
photoautotrophs, 192
physics, 175, 179
Piantanida, Nicholas, 111–12
pioneers, 128–29
Plato, 63
Pleistocene, 169
poetics, 148–49
poetic thought, 44
political, realm of the, 34
polyvinyl pyrrolidone K12, 139
populous solitude, 71
postanthropocentrism, 39; knowledge
    production, 174

thinking/thought (continued)
69, 73, 75, 78–79, 149–50, 188, 214; pow-
er(s) of, 28; proprietary, 31; questioning, 18;
representational, 28; romanticization of, 19;
State, 44, 69, 142; temporality of, 22
think-with, 48, 56, 159, 168; Judith Butler, 72;
Donna Haraway, 40–43; movement, 66;
transhumanism, 94
Thirty Years' War, 139
Thompson, Ken, 90
thoughtfulness, 20
thoughtlessness, 18–19
topology, political, 151
totalitarianism, 137
traitorism, 17, 40, 70, 163, 206, 218
transcendence, 79–80, 169; expansion, 21
transhumanism, 21–24, 38, 81–82, 86–88, 198–
201; goals, 85; mode of sensemaking, 106;
normativity, 88–95; relationality, 81; thought,
78–79, 81, 84–85, 93, 108, 139; transcending
death, 178–79. See also posthumanism
trickster, 17
Trinity, 33
Trump, Donald, 106
trust, 164
truths, 15, 21, 66, 68, 158; multiple, 62
Tsing, Anna, 27, 130
Turner, Jackson, 130

übermensch, 80
undercommons, 23
Union Carbide, 198
United Nations, 161; Convention on the Law
of the Seas (UNCLOS), 130, 145, 155
United States, 106, 130, 155, 207–8, 212
Universal Declaration of Animal Rights, 195
university, 62–63
unknown, absolute, 72–73
unpersoning, 191
unrepresentability, 79
US Constitution, 138
US National Aeronautic Association, 141
US Supreme Court, 171

Valentín, Felix, 90
vegetal model of thought, 23
Venter, Craig, 173
Vescovo, Victor L., 161

viewing, 36
violence, 7, 25, 33, 70, 72, 74, 170,
183, 200, 206
Virgin Galactic, 136
Viveiros de Castro, Eduardo, 2, 5–6
void, the, 38, 41, 52, 54, 66
von Kármán line, 141

Waldron, Jeremy, 20
Walter, Ulrich, 114
wars, 23, 29
Watson, Julia, 188
Web 3.0, 124
Weil, Kari, 196
Weizman, Eyal, 162–63
Wellerstein, Alex, 117
West Africa, 162
Western: epistemologies, 197; feminism, 38;
legal thought, 157; liberal thought, 154;
philosophy, 7
Western Sahara, 162
Whitehead, Alfred North, 38–39, 58–66, 75
Wilczek, Frank, 203
will, 132
Williams, Patricia J., 78, 80, 207
Wilson, Shawn, 40
withdrawing, 28
Wolf, Christa, 57, 70, 72
Wolfe, Cary, 86
Woodmansee, Martha, 181
World Health Organization, 106, 212
world-thinking, 78

X, 105, 122, 125
X-1000, 139
xenon, 149
XI Protocol, 124

Yang, Neon, 16
Yin-Yang, 82
YouTube, 124

Z-1000, 139
Zalik, Anna, 161
ZEDES (Zones for Economic Development
and Employment), 90–93
Zhang, Laura, 136
Zimmerman, George, 208

www.ingramcontent.com/pod-product-compliance
Lightning Source LLC
Chambersburg PA
CBHW032344280326
41935CB00008B/450